SpringerBriefs in Statistics

SpringerBriefs present concise summaries of cutting-edge research and practical applications across a wide spectrum of fields. Featuring compact volumes of 50 to 125 pages, the series covers a range of content from professional to academic. Typical topics might include:

- A timely report of state-of-the art analytical techniques
- A bridge between new research results, as published in journal articles, and a contextual literature review
- A snapshot of a hot or emerging topic
- An in-depth case study or clinical example
- A presentation of core concepts that students must understand in order to make independent contributions

SpringerBriefs in Statistics showcase emerging theory, empirical research, and practical application in Statistics from a global author community.

SpringerBriefs are characterized by fast, global electronic dissemination, standard publishing contracts, standardized manuscript preparation and formatting guidelines, and expedited production schedules.

Janis E. Johnston • Kenneth J. Berry •
Michael A. Long

Permutation Statistical Methods: Measures of Relationship

Springer

Janis E. Johnston
Colorado State University
Fort Collins, CO, USA

Kenneth J. Berry
Colorado State University
Fort Collins, CO, USA

Michael A. Long
Oklahoma State University
Stillwater, OK, USA

ISSN 2191-544X ISSN 2191-5458 (electronic)
SpringerBriefs in Statistics
ISBN 978-3-032-26607-1 ISBN 978-3-032-26608-8 (eBook)
https://doi.org/10.1007/978-3-032-26608-8

This Springer imprint is published by the registered company Springer Nature Switzerland AG
The registered company address is: Gewerbestrasse 11, 6330 Cham, Switzerland

For Dr. Paul W. Mielke, Jr. (1931–2019),
teacher, colleague, collaborator, and friend.

Preface

This Springer Brief on *Permutation Statistical Methods: Measures of Relationship* is a companion volume by the authors to *Permutation Statistical Methods: Test of Differences*. While the earlier volume concentrated on statistical tests of differences, including tests of differences for one-sample tests, two independent samples, two matched samples, multiple independent samples, and multiple matched samples, the present volume concentrates on measures of relationship, including measures of nominal-nominal association, ordinal-ordinal association, interval-interval correlation and agreement, nominal-ordinal association, nominal-interval association, and ordinal-interval association.

This book is intended to offer a short introduction to permutation statistical methods for analyzing correlation, association, and agreement. Because this is a long-overlooked area of statistical inference, however, some readers might find it useful as a methodological reference. Throughout the book the emphasis is on permutation statistical methods, although the results of the permutation analyses are always compared and contrasted with the results of conventional statistical analyses, with which the reader is expected to be familiar. It is assumed that the reader has a background in elementary social science statistics, such as might be taught in departments of sociology, criminal justice, economics, political science, or psychology.

The book applies a basic approach to statistical analyses based on permutations of observed datasets that avoid many of the rigorous assumptions underlying conventional statistical methods used in conventional statistics, such as taught in most introductory courses.

Two models of statistical inference are considered in this book: the Neyman–Pearson population model and the Fisher–Pitman permutation model. The Neyman–Pearson population model of statistical inference is the standard model taught in all introductory classes and will be familiar to most readers. The Neyman–Pearson population model is named after contemporaries Jerzy Neyman (1894–1981) and Egon Pearson (1895–1980). The model was designed to make inferences about population parameters and provide approximate probability values. The Fisher–Pitman permutation model is less well known than the Neyman–Pearson population model.

The Fisher–Pitman permutation model is named after R.A. Fisher (1890–1962) and E.J.G. Pitman (1897–1993) and includes two different, but related, permutation methodologies: exact permutation methods and Monte Carlo permutation methods.

While the Neyman–Pearson population model needs little or no introduction, the Fisher–Pitman permutation model of statistical inference is less likely to be familiar to many readers. In contrast to conventional statistical tests based on the Neyman–Pearson population model, tests based on the Fisher–Pitman permutation model are distribution-free, entirely data-dependent, appropriate for nonrandom samples, provide exact probability values, and are ideal for small datasets. On the other hand, permutation statistical methods can be computationally intensive, often requiring many thousands of calculations.

For the Fisher–Pitman permutation model, exact and Monte Carlo permutation methods are described and compared. Exact permutation tests analyze all possible arrangements of the observed data and provide an exact probability value, as the name implies. Exact methods often result in a very large number of arrangements. For example, two small independent samples of size $n_1 = n_2 = 15$ generate 155,117,520 possible arrangements of the observed data. When the number of arrangements becomes very large, Monte Carlo permutation methods analyze a large random sample of arrangements, providing an approximate, but highly accurate, probability value. With a sufficient number of randomly selected samples, a Monte Carlo probability value can be computed to any reasonable accuracy.

While computer programs for the various permutation analyses are not included in this book, there are a variety of available programs for both exact and Monte Carlo analyses available. One source is *Statistical Permutation Methods with R* by Berry, Kvamme, Johnston, and Mielke. R is a popular open-source, free programming language.

In addition, RStudio offers a GUI-like interface for R. RStudio is a free software add-on that provides an integrated development environment, which not only facilitates general usage of R but specifically aims at code building and software development.

Fort Collins, CO, USA — Janis E. Johnston
Fort Collins, CO, USA — Kenneth J. Berry
Stillwater, OK, USA — Michael A. Long
November 2025

Competing Interests The authors have no competing interests to declare that are relevant to the content of this manuscript.

Contents

Chapter 1
Introduction

Abstract This chapter of *Permutation Statistical Methods: Measures of Relationship* provides an introduction to the next five chapters. Brief overviews of each chapter are provided with emphasis on nominal-nominal, ordinal-ordinal, interval-interval, nominal-ordinal, nominal-interval, and ordinal-interval measures of relationship.

Keywords Permutation methods · Measurement levels · Nominal association · Ordinal association

The focus of this Springer Brief is on a permutation approach to the measurement of relationships between and among variables at various levels of measurement. While a large number of methods exist for measuring the magnitude of association between two variables, there is considerable difficulty in interpreting and comparing the various measures, as they often differ in structure, logic, and interpretation. Permutation statistical methods bring a common platform to the measures without the usual assumptions of normality or random sampling.

The organization of the book is based on three levels of measurement: nominal, ordinal, and interval, as introduced by Stevens in [3]. As Leik and Gove noted, as data progress from nominal to ordinal to interval, measures of relationship should ideally incorporate the added properties of the level of measurement into the logic used at the previous level [2, p. 279]. Unfortunately, this is not the case with the numerous and diverse measures of association currently in use. Thus the various measures of relationship developed over the years constitute a hodgepodge of approaches, logic, and interpretation. In examining the various measures of relationship, it is convenient to categorize the measures by the level of measurement for which they are most appropriate, recognizing that some measures are suitable for more than one level of measurement.

Besides structure and interpretation, a major drawback to measures of relationship lies in determining the probability of the obtained measure under the null hypothesis. There are two major approaches to determining probability: the Neyman–Pearson population model and the Fisher–Pitman permutation model [1].

J. E. Johnston et al., *Permutation Statistical Methods: Measures of Relationship*, SpringerBriefs in Statistics, https://doi.org/10.1007/978-3-032-26608-8_1

The Neyman–Pearson population model is named after contemporaries Jerzy Neyman (1894–1981) and Egon Pearson (1895–1980). The model was designed to make inferences about population parameters and provide approximate probability values. The Fisher–Pitman permutation model is named after R.A. Fisher (1890–1962) and E.J.G. Pitman (1897–1993) and includes two different, but related, permutation methodologies: exact permutation methods and Monte Carlo permutation methods.

The Neyman–Pearson population model requires several assumptions that are seldom satisfied in practice and are often inappropriate for the lower levels of measurement, e.g., independence, random sampling from a parent population, an underlying Gaussian distribution (often referred to as the "normal" distribution) for the target variable in the population, and homogeneity of variance (and covariance, when appropriate). In this book, the permutation model is used exclusively as it is free of any distributional assumptions, does not require random sampling, and is ideally suited for the analysis of small samples.

1.1 Overviews of Chaps. 2–6

Brief summaries of the substantive chapters in the book are provided in this section. The summaries are simply intended to provide overviews of the various chapters and are not meant to provide detailed summaries of the chapters. In the age of electronic books, oftentimes a single chapter is downloaded by a user. Consequently, in this book, each chapter is designed to be self-contained. Therefore, there is a small amount of redundancy among the chapters.

1.1.1 Chapter 2: Permutation Statistical Methods

Chapter 2 presents a brief overview of permutation statistical methods as applied to measures of relationship. Permutation methods generally consist of two types: exact and Monte Carlo methods. For exact permutation measures, a test statistic is computed on a dataset, all possible arrangements of the data are identified, and the test statistic is computed for each arrangement. The proportion of arrangements with test statistic values as or more extreme than the value of the test statistic computed on the original data is the exact probability value. The number of possible arrangements can become quite large even for small datasets. One alternative is a Monte Carlo permutation procedure in which a large, but comparatively small, subset of all possible arrangements of the data is examined. The approximate probability value is the proportion of arrangements in the subset with test statistic values that are as or more extreme than the test statistic value computed on the original data arrangement. Generally, a Monte Carlo probability value can be computed to any accuracy level by simply increasing the number of arrangements in the sample, e.g., for most data, $n = 10{,}000$ data arrangements provide a probability value with two decimal places of accuracy, and increasing the sample size by two orders of magnitude provides an additional place of accuracy, e.g., $n = 1{,}000{,}000$ results in a probability with three decimal places of accuracy.

1.1.2 Chapter 3: Measures of Nominal Association

Measures of relationship between two nominal-level variables are common in the statistical literature. Nominal-level variables, also known as categorical variables, are used to classify objects without implying any inherent order, for example, marital status, gender, political affiliation, or religious preference. Chapter 3 presents exact and Monte Carlo permutation statistical methods applied to two cross-classified nominal-level variables. Included in Chap. 3 are Pearson's chi-squared test statistic, a discussion of standardized residuals, and measures of association based on Pearson's chi-squared statistic; specifically, Pearson's ϕ^2 and Cramér's V^2. Fisher's exact permutation test for a 2×2 cross-classification is presented using an efficient recursion technique. Goodman and Kruskal's t_a and t_b asymmetric measures of nominal association are described and illustrated.

1.1.3 Chapter 4: Measures of Ordinal Association

Measures of relationship between two ordinal-level variables, such as birth order, socio-economic status, academic rank, or Likert scores[1] are widely used in the physical and social sciences. Chapter 4 presents exact and Monte Carlo permutation statistical methods applied to two cross-classified ordinal variables. The first section provides a brief introduction to permutation statistical methods for ordinal data. Kendall's S test statistic, Kendall's τ_a measure, Kendall's τ_b measure, and Stuart's τ_c measure are described and illustrated. Goodman and Kruskal's γ symmetric measure is presented, and Somers' d_{yx} and d_{xy} asymmetric measures are introduced. Finally, a new procedure for determining the maximum value of Kendall's S is described.

1.1.4 Chapter 5: Measures of Correlation

Measures of relationship between two interval-level variables, such as family income, years of education, farm size in acres, or populations of cities, are ubiquitous in the popular and academic literatures. Chapter 5 presents exact and

[1] Likert scores are typical in survey research, for example, when a respondent is asked to identify level of disagreement/agreement with a statement, ranked on a score of 1–5, where 1 represents strong disagreement and 5 represents strong agreement.

Monte Carlo permutation statistical methods applied to two correlated variables. Permutation statistical methods as applied to Pearson's conventional product-moment correlation coefficient, to the point–biserial correlation coefficient, and to the biserial correlation coefficient are described and illustrated. Data analyses employing the intraclass correlation coefficient are presented. Measures of association based on Spearman's rank-order correlation coefficient are described and illustrated, and data analyses employing Spearman's footrule measure of inter-rater agreement are presented. Finally, permutation methods as applied to Kendall and Babington Smith's coefficient of concordance for multiple samples are introduced.

1.1.5 Chapter 6: Mixed Measures of Association

Measures of relationship between two mixed variables (e.g., nominal-ordinal pairs), such as gender and academic rank, gender and age, or marital status and age, are less common in statistical literatures, but play an important role in selected analyses. Chapter 6 applies exact and Monte Carlo permutation statistical methods to measures of association designed for two variables at different levels of measurement: permutation statistical methods for measuring association between a nominal-independent variable and an ordinal-dependent variable, between a dichotomous nominal variable and an ordinal variable, between a nominal variable and an interval variable, and between an ordinal variable and an interval variable. Freeman's θ measure is described and illustrated as a measure of nominal-ordinal association. Cureton's r_{rb} correlation coefficient, the point-biserial correlation coefficient, and the biserial correlation coefficient are described and illustrated as measures of nominal-interval association. Jaspen's multiserial correlation coefficient is described and illustrated as a measure of ordinal–interval association.

References

1. K.J. Berry, P.W. Mielke, J.E. Johnston, *Permutation Statistical Methods: An Integrated Approach* (Springer, Cham CH, 2016)
2. R.K. Leik, W.R. Gove, Integrated approach to measuring association, in *Sociological Methodology*, ed. by H.L. Costner (Jossey Bass, San Francisco, CA, 1971), pp. 279–301
3. S.S. Stevens, On the theory of scales and measurement. Science **103**, 667–680 (1946)

Chapter 2
Permutation Statistical Methods

Abstract This chapter presents an overview of exact and Monte Carlo permutation statistical methods. Five sections describe and illustrate the basics of permutation statistical methods as applied to measures of relationship. The first section simply provides an introduction to the chapter. The second section provides an introduction to permutation statistical methods. The third section details the differences between tests of differences and measures of relationships. The fourth section describes exact and Monte Carlo permutation statistical methods for tests of differences. The fifth section describes exact and Monte Carlo permutation statistical methods for measures of association and correlation. The sixth section describes some advantages and disadvantages of permutation statistical tests and measures.

Keywords Exact tests · Monte Carlo methods · Probability · Recursion

2.1 Introduction

This chapter presents a brief overview of permutation statistical methods as applied to measures of relationship. Exact and Monte Carlo methods are the most common types of permutation methods. In an exact permutation test, a test statistic is computed on data associated with a collection of objects, all possible arrangements of the objects are generated, and a test statistic is computed for each arrangement of the objects. The null hypothesis is that each arrangement of objects is equally likely. The proportion of arrangements with test statistic values as or more extreme than the value of the test statistic computed on the original arrangement of the data is the exact probability value. The number of possible arrangements can become quite large even for small datasets. For example, an experimental design with $n = 30$ objects randomized into three treatments with $n_1 = 9$, $n_2 = 10$, and $n_3 = 11$ objects in each treatment requires

$$M = \frac{n!}{n_1!\, n_2!\, n_3!} = \frac{30!}{9!\ 10!\ 11!} = 5{,}046{,}360{,}719{,}400$$

J. E. Johnston et al., *Permutation Statistical Methods: Measures of Relationship*, SpringerBriefs in Statistics, https://doi.org/10.1007/978-3-032-26608-8_2

arrangements of the data. Therefore, alternative methods are often needed to approximate the exact probability value.

One alternative is a Monte Carlo permutation procedure in which a large subset of all possible arrangements of the data is examined. The usual method employed in a Monte Carlo permutation procedure is to repeatedly shuffle the observed sequence of n data points and to select a random subset of all possible arrangements of the data. The proportion of arrangements in the subset with test statistic values as or more extreme than the value of the test statistic computed on the original arrangement of the data is the approximate probability value. Provided that the exact probability value is not too small and that the number of shuffles is large, Monte Carlo permutation procedures provide excellent approximations to exact probability values.

2.2 Permutation Statistical Methods

The permutation model was introduced by R.A. Fisher in [14], further developed by R.C. Geary in [16] and T. Eden and F. Yates in [6], and made explicit in three seminal articles by E.J.G. Pitman in [44–46]. In a permutation statistical test, the only assumption is that experimental variability caused the observed result.

Permutation statistical tests are considered by many to be a gold standard against which conventional statistical tests should be evaluated and validated. In 1940, Friedman, comparing tests of significance for multiple rankings, referred to an exact permutation test as "the correct one" [15, p. 88]. In 1973, Feinstein remarked that conventional statistical tests "yield reasonably reliable approximations of the more exact results provided by permutation procedures" [12, p. 912]. In 1992, Good noted that Fisher regarded randomization as a technique for validating tests of significance, ensuring that conventional probability values were accurate [17, p. 263]. Bakeman, Robinson, and Quera remarked in 1996 that "like Read and Cressie…, we think permutation tests represent the standard against which asymptotic tests must be judged" [1, p. 6]. And in 2007, Edgington and Onghena observed that "randomization tests…have come to be recognized by many…as the 'gold standard' of statistical tests for randomized experiments" [11, p. 9].[1]

The value of permutation statistical tests was recognized by early statisticians, even during periods in which the computationally intensive nature of permutation tests made them impractical. In 1955, Kempthorne wrote that "tests of significance in the randomized experiment have frequently been presented by way of normal law theory, whereas their validity stems from randomization theory" [25, p. 947] and "there seems little point in the present state of knowledge in using [a] method of inference other than randomization analysis" [25, p. 966]. Similarly, in 1959, Scheffé stated that the conventional analysis of variance F ratio "can often be

[1] In the literature, the terms "permutation" and "randomization" are often used interchangeably.

regarded as a good approximation to a permutation test, which is an exact test under a less restrictive model" [47, p. 313]. In 1966, Kempthorne reemphasized that "the proper way to make tests of significance in the simple randomized experiments [sic] is by way of the randomization (or permutation) test" [26, p. 20] and "in the randomized experiment one should, logically, make tests of significance by way of the randomization test" [26, p. 21]. Later, in 1968, Bradley observed that "eminent statisticians have stated that the randomization test is the truly correct one and that the corresponding parametric test is valid only to the extent that it results in the same statistical decision" [4, p. 85].

Because permutation statistical methods are inherently computationally intensive, it took the development of high-speed computing for permutation methods to achieve their potential. Today, a small laptop computer outperforms even the largest mainframe computers of previous decades [3, p. 4]. Consequently, in the twenty-first century, permutation statistical methods have become both feasible and practical and have found applications in diverse fields of research ranging from agriculture to zoology. Fields of research that examine small nonrandom samples, such as atmospheric science, psychology, ecology, biology, and medicine, have been especially receptive to permutation methods. This is due in large part to strong advocates of permutation statistical methods in these fields, including Hugh Dudley [32–35], Eugene Edgington [7–10], Alvan Feinstein [12, 13], Phillip Good [18–21], Oscar Kempthorne [24–27], John Ludbrook [28–31], Bryan Manly [36–39], Paul Mielke [40–43], and John Tukey [5, 48–50].

2.3 Differences and Relationships

Many introductory textbooks and statistics teachers divide statistical analyses into two types of problems for pedagogical purposes. One problem type tests for differences between or among population means. Examples of such tests of differences are Student's two-sample t test and various analysis of variance tests. The second problem type measures relationships between or among population variables. Examples of such measures of relationship are Pearson's product-moment correlation coefficient and various advanced measures based on multiple correlation or regression.

2.4 Tests of Differences

Permutation tests of differences come in two forms: exact and Monte Carlo. Exact permutation statistical tests evaluate all arrangements of the observed data and deliver an exact probability value. Monte Carlo permutation tests evaluate a large random sample of arrangements of the observed data and deliver an approximate probability value.

2.4.1 An Exact Permutation Test

For an illustration of an exact permutation test of differences, consider a test between two population means. Table 2.1 lists the per capita relief expenditures in 1831, in shillings, for $n = 22$ parishes (identified only by number) in two counties: Southampton and Suffolk.

A compulsory system of poor relief was instituted in England during the reign of Elizabeth I (1558–1603). Essentially, the laws distinguished three major categories of dependents: the vagrant, the involuntary unemployed, and the helpless. Although the role played by poor relief was significantly modified by the Poor Law Amendment Act of 1834, the Crusade Against Outrelief of the 1870s, and the adoption of various social insurance programs in the early twentieth century, the Poor Law continued to assist the poor until it was replaced by the welfare state in 1948.

In 1831, Southampton county consisted of $n_1 = 15$ parishes with a mean relief expenditure of $\bar{x}_1 = 15.4171$ shillings and a sample standard deviation of $s_1 = 7.4081$ shillings. Suffolk county consisted of $n_2 = 7$ parishes with a mean relief expenditure of $\bar{x}_2 = 22.4808$ shillings and a sample standard deviation of $s_2 = 7.0321$ shillings.

2.4.2 Asymptotic Probability

For the Southampton and Suffolk county relief data in Table 2.1, a conventional Student's two-sample t test yields $t = -2.1147$, and with $n_1 + n_2 - 2 = 15 + 7 - 2 = 20$ degrees of freedom the asymptotic two-tail probability value under the null hypothesis is $P = 0.0472$.

2.4.3 Permutation Probability

There are only

$$M = \frac{(n_1 + n_2)!}{n_1!\, n_2!} = \frac{(15 + 7)!}{15!\; 7!} = 170{,}544$$

arrangements of the Southampton and Suffolk county relief data, making an exact permutation analysis feasible. Exactly 9,010 t values in the reference set of all

Table 2.1 Average per capita relief expenditures for Southampton and Suffolk counties in shillings: 1831

Southampton		Suffolk	
Parish	Relief	Parish	Relief
1	6.731808	6	26.383673
6	16.156615	10	16.727664
7	14.760218	11	27.628032
8	15.057353	13	19.914255
12	11.001482	19	13.833671
15	29.089955	24	33.827534
18	11.818136	28	19.050737
25	16.002180		
27	18.761256		
29	32.443278		
31	15.447992		
36	15.756267		
38	4.257547		
39	8.611310		
40	15.361136		

Source Table 2.1 is adapted from Johnston, Berry, Mielke: Quantitative historical methods: A permutation alternative [23, p. 37]

possible t values are equal to or more extreme than the observed value of $t = -2.1147$, yielding an exact probability value of

$$P(t \geq t_{\text{o}} \,|\, \text{H}_0) = \frac{\text{number of } t \text{ values } \geq t_{\text{o}}}{M} = \frac{9{,}010}{170{,}544} = 0.0528 \;,$$

where t_{o} is the observed value of t. For this example analysis, the difference between the exact and asymptotic probability values is only $P(0.0528 - 0.0472) = 0.0056$.

2.4.4 A Monte Carlo Permutation Test

To illustrate a Monte Carlo permutation test of differences, consider Table 2.2, which contains the per capita relief expenditures in 1831, in shillings, for $n = 36$ parishes (identified only by number) in two counties: Oxford and Hertford. In 1831, Oxford county consisted of $n_1 = 24$ parishes with a mean relief expenditure of $\bar{x}_1 = 20.2766$ shillings and a sample standard deviation of $s_1 = 7.6408$ shillings. Hertford county consisted of $n_2 = 12$ parishes with a mean relief expenditure of $\bar{x}_2 = 13.4720$ shillings and a sample standard deviation of $s_2 = 6.1270$ shillings.

Table 2.2 Average per capita relief expenditures for Oxford and Hertford counties in shillings: 1831

Oxford		Hertford	
Parish	Relief	Parish	Relief
1	20.361860	2	27.974783
2	29.086095	4	6.417284
5	14.931757	7	10.484120
8	24.123211	11	10.005750
10	18.207501	13	9.769865
11	20.728732	14	15.866521
12	8.119472	15	19.342360
13	14.020071	17	17.145218
17	18.424789	20	13.134206
18	34.546600	21	10.041964
19	16.092713	22	15.083824
22	24.616592	27	6.398451
23	25.468298		
24	12.563194		
29	13.278003		
31	27.302973		
34	29.605508		
36	13.613192		
39	11.371418		
45	21.524807		
49	20.940801		
52	11.595229		
55	18.235469		
56	37.880889		

Source Table 2.2 is adapted from Johnston, Berry, Mielke: Quantitative historical methods: A permutation alternative [23, p. 37]

2.4.5 *Asymptotic Probability*

For the Oxford and Hertford county relief data in Table 2.2, a conventional Student's two-sample t test yields $t = +2.6783$, and with $n_1 + n_2 - 2 = 24 + 12 - 2 = 34$ degrees of freedom, the asymptotic two-tail probability value under the null hypothesis is $P = 0.0113$.

2.4.6 *Permutation Probability*

There are

$$M = \frac{(n_1 + n_2)!}{n_1!\, n_2!} = \frac{(24 + 12)!}{24!\ 12!} = 1{,}251{,}677{,}700$$

arrangements of the Oxford and Hertford county relief data, making an exact permutation analysis impractical. Based on $L = 1{,}000{,}000$ random arrangements of the observed data[2] with $n_1 = 24$ and $n_2 = 12$ preserved for each arrangement, 8,478 of the calculated t values are equal to or more extreme than $t = +2.6783$, yielding a Monte Carlo exact probability value of

$$P(t \geq t_{\text{o}} \mid \text{H}_0) = \frac{\text{number of } t \text{ values } \geq t_{\text{o}}}{L} = \frac{8{,}478}{1{,}000{,}000} = 0.0085 ,$$

where t_{o} is the observed value of t. For this example analysis, the difference between the Monte Carlo and asymptotic probability values is only $P(0.0113 - 0.0085) = 0.0028$.

2.5 Measures of Relationship

To illustrate a Monte Carlo permutation measure of relationship, consider Table 2.3, which lists a husband's occupational prestige score (x) and a family's social status score (y) for $n = 10$ families. The data are hypothetical. The number of observations is kept small for illustration purposes.

For the bivariate data in Table 2.3, $\sum_{i=1}^{n} x_i = 420$, $\sum_{i=1}^{n} x_i^2 = 22{,}600$, $\sum_{i=1}^{n} y_i = 500$, $\sum_{i=1}^{n} y_i^2 = 32{,}606$, $\sum_{i=1}^{n} x_i y_i = 25{,}640$, and Pearson's product-moment correlation coefficient is

Table 2.3 Husband's occupational prestige (x) and family's social status (y)

Family	x	y
A	75	80
B	70	87
C	60	91
D	55	44
E	50	22
F	40	58
G	25	52
H	20	10
I	15	38
J	10	18

[2] In general, $L = 1{,}000{,}000$ random arrangements provide a probability value with three decimal places of accuracy [22].

$$r_{xy} = \frac{n\sum_{i=1}^{n} x_i y_i - \sum_{i=1}^{n} x_i \sum_{i=1}^{n} y_i}{\sqrt{\left[n\sum_{i=1}^{n} x_i^2 - \left(\sum_{i=1}^{n} x_i\right)^2\right]\left[n\sum_{i=1}^{n} y_i^2 - \left(\sum_{i=1}^{n} y_i\right)^2\right]}}$$

$$= \frac{(10)(25{,}640) - (420)(500)}{\sqrt{[(10)(22{,}600) - (420)^2][(10)(32{,}606) - (500)^2]}} = +0.7554 \, .$$

2.5.1 Asymptotic Probability

For the occupational prestige (x) and social status (y) data in Table 2.3, a conventional Student's t test for r_{xy} is

$$t = r_{xy}\sqrt{\frac{n-2}{1-r_{xy}^2}} = +0.7554\sqrt{\frac{10-2}{1-(0.7554)^2}} = +3.2610,$$

and with $n-2 = 10-2 = 8$ degrees of freedom, the asymptotic two-tail probability value under the null hypothesis is $P = 0.0115$.

2.5.2 Monte Carlo Probability

Because there are

$$M = n! = 10! = 3{,}628{,}800$$

possible arrangements of the bivariate data in Table 2.3, an exact permutation test may be impractical and a Monte Carlo test is advised. Based on $L = 1{,}000{,}000$ random arrangements of the observed data under the null hypothesis, there are 15,600 r_{xy} values equal to or greater than $r_{xy} = +0.7554$, yielding a Monte Carlo probability value of

$$P(r_{xy}\,|\,\mathrm{H}_0) = \frac{\text{number of } r_{xy} \text{ values} \geq r_\text{o}}{L} = \frac{15{,}600}{1{,}000{,}000} = 0.0156 \, ,$$

where r_o is the observed value of Pearson's r_{xy}.

2.5.3 An Exact Probability

While an exact permutation test is not practical for the data in Table 2.3, it is not impossible. There are $M = 3{,}628{,}800$ possible arrangements and 60,571 r_{xy} values greater than $r_{xy} = +0.7554$. Thus, for the bivariate data in Table 2.3, the exact probability of $r_{xy} = +0.7554$ is

$$P(r_{xy}\,|\,\mathrm{H}_0) = \frac{\text{number of } r_{xy} \text{ values } \geq r_\mathrm{o}}{M} = \frac{60{,}571}{3{,}628{,}800} = 0.0167\ .$$

2.5.4 Probability Comparisons

For this example analysis, the differences among the three probability values are small. The difference between the asymptotic and Monte Carlo probability values is only $P(0.0156-0.0115) = 0.0041$, the difference between the asymptotic and exact probability values is only $P(0.0167-0.0115) = 0.0052$, and the difference between the Monte Carlo and exact probability values is only $P(0.0167-0.0156) = 0.0011$.

2.6 Advantages of Permutation Methods

Compared with conventional asymptotic methods, permutation statistical methods do not assume normality, homogeneity of variance, or random sampling. In addition, permutation methods are data dependent in that all the information required for analysis is contained within the observed data. On the other hand, permutation methods can be computationally intensive [2, p. 21].

Alvan Feinstein was a strong advocate for permutation methods. Trained as a medical doctor, Feinstein is widely regarded as the founder of clinical epidemiology and patient-oriented medicine and the originator of clinimetrics: the application of mathematics to the field of medicine. In 1973, Feinstein published a formative article on "The role of randomization in sampling, testing, allocation, and credulous idolatry" [12].

Writing for a statistically unsophisticated readership, Feinstein distinguished between socio-political research where the purpose was usually to estimate a population parameter and medical research where the purpose was typically to contrast a difference between two groups.

Feinstein put forth some advantages and disadvantages of permutation tests that were insightful for the time and foreshadowed later research. In terms of permutation tests, he listed five advantages:

1. The result of a permutation test is a direct, exact probability value for the random likelihood of the observed difference.
2. Permutation tests do not require any unwarranted inferential estimations of means, variances, pooled variances, or other parameters of an unobserved, hypothetical parent population. The tests are based solely on the evidence that was actually obtained.[3]
3. The investigator is not forced into making any erroneous assumptions either that the contrasted groups were chosen as random samples from a parent population, or that treatments under study were randomly allocated to the two groups.
4. The investigator is not forced into making any erroneous or unconfirmable assumptions about a Gaussian (or any other) distribution for the parent population, or about equal variances in the contrasted groups.
5. A permutation test can be applied to groups of any size, no matter how large or small. There are no degrees of freedom to be considered. In the case of a contingency table, there is no need to worry about the magnitude of the expected value, no need to calculate expectations based on fractions of people, and no need to worry about applying, or not applying, Yates correction for continuity.

Feinstein observed that while there were definite advantages to permutation tests, there were also disadvantages. The first three (of four) he considered as features that contributed to "the existing state of statistical desuetude" and labeled them inertia, ideology, and information [12, p. 911]:

1. Inertia: It is easier for many teachers to continue the inertia of teaching what they were taught years ago than to revise the contents of their lectures.
2. Ideology: Investigators who ideologically believe that the goal of science is to estimate parameters and variances will have no enthusiasm for tests that do not include or rely on these estimations.
3. Information: Many investigators have a deep-seated horror of doing anything that might entail losing information.
4. Permutation tests are notoriously difficult to calculate.[4]

2.7 Summary

Chapter 2 presented a brief overview of permutation statistical methods with an emphasis on exact and Monte Carlo methods under the Fisher–Pitman permutation model. The chapter explained that exact permutation methods (1) define a suitable test statistic, (2) generate all possible arrangements of the observed data, (3) compute the selected test statistic for each arrangement, and (4) compute the probability as the proportion of generated values equal to or more extreme than the observed value. The chapter also explained that Monte Carlo permutation methods (1) define a suitable test statistic, (2) generate a large random sample of all possible arrangements, (3) compute the selected test statistic for each random arrangement,

---

[3] In this second advantage, Feinstein clearly described the data-dependent nature of permutation tests, anticipating by many years later research on permutation methods.

[4] Note that this fourth disadvantage was published in 1971 when high-speed computing was not widely available.

and (4) compute the probability as the proportion of random values equal to or more extreme than the observed value. Measures of relationship were described and illustrated with both exact and Monte Carlo analyses.

References

1. R. Bakeman, B.F. Robinson, V. Quera, Testing sequential association: estimating exact p values using sampled permutations. Psychol. Meth. **1**, 4–15 (1996)
2. K.J. Berry, J.E. Johnston, P.W. Mielke, *The Measurement of Association: A Permutation Statistical Approach* (Springer, Cham CH, 2018)
3. K.J. Berry, P.W. Mielke, J.E. Johnston, *Permutation Statistical Methods: An Integrated Approach* (Springer, Cham, CH, 2016)
4. J.V. Bradley, *Distribution-Free Statistical Tests* (Prentice–Hall, Englewood Cliffs, NJ, 1968)
5. D.R. Brillinger, L.V. Jones, J.W. Tukey, The role of statistics in weather resources management, in *The Management of Weather Resources II* (United States Government Printing Office, Washington, DC, 1978)
6. T. Eden, F. Yates, On the validity of Fisher's z test when applied to an actual example of non-normal data. J. Agric. Sci. **23**, 6–17 (1933)
7. E.S. Edgington, Randomization tests. J. Psychol. **57**, 445–449 (1964)
8. E.S. Edgington, Statistical inference and nonrandom samples. Psychol. Bull. **66**, 485–487 (1966)
9. E.S. Edgington, Approximate randomization tests. J. Psychol. **72**, 143–149 (1969)
10. E.S. Edgington, *Randomization Tests* (Marcel Dekker, New York, 1980)
11. E.S. Edgington, P. Onghena, *Randomization Tests*, 4th edn. (Chapman & Hall/CRC, Boca Raton, FL, 2007)
12. A.R. Feinstein, Clinical Biostatistics XXIII: the role of randomization in sampling, testing, allocation, and credulous idolatry (Part 2). Clin. Pharmacol. Ther. **14**, 898–915 (1973)
13. A.R. Feinstein, *Clinical Biostatistics* (C.V. Mosby, St. Louis, 1977)
14. R.A. Fisher, *Statistical Methods for Research Workers* (Oliver and Boyd, Edinburgh, 1925)
15. M. Friedman, A comparison of alternative tests of significance for the problem of m rankings. Ann. Math. Stat. **11**, 86–92 (1940)
16. R.C. Geary, Some properties of correlation and regression in a limited universe. Metron **7**, 83–119 (1927)
17. I.J. Good, Further comments concerning the lady tasting tea or beer: P-values and restricted randomization. J. Stat. Comput. Sim. **40**, 263–267 (1992)
18. P.I. Good, *Permutation Tests: A Practical Guide to Resampling Methods for Testing Hypotheses* (Springer, New York, 1994)
19. P.I. Good, Permutation, *Parametric and Bootstrap Tests of Hypotheses* (Springer, New York, 1994)
20. P.I. Good, *Resampling Methods: A Practical Guide to Data Analysis* (Birkhäuser, Boston MA, 1999)
21. P.I. Good, *Permutation Tests: A Practical Guide to Resampling Methods for Testing Hypotheses*, 2nd edn. (Springer, New York, 2000)
22. J.E. Johnston, K.J. Berry, P.W. Mielke, Permutation tests: precision in estimating probability values. Perc. Motor Skills, **105**, 915–920 (2007)
23. J.E. Johnston, K.J. Berry, P.W. Mielke, Quantitative historical methods: a permutation alternative. Historical Methods **42**, 35–39 (2009)
24. O. Kempthorne, *The Design and Analysis of Experiments* (Wiley, New York, 1952)
25. O. Kempthorne, The randomization theory of experimental inference. J. Am. Stat. Assoc. **50**, 946–967 (1955)

26. O. Kempthorne, Some aspects of experimental inference. J. Am. Stat. Assoc. **61**, 11–34 (1966)
27. O. Kempthorne, Why randomize? J. Stat. Plan. Infer. **1**, 1–25 (1977)
28. J. Ludbrook, Advantages of permutation (randomization) tests in clinical and experimental pharmacology and physiology. Clin. Exp. Pharmacol. Physiol. **21**, 673–686 (1994)
29. J. Ludbrook, Issues in biomedical statistics: comparing means by computer intensive tests. Aust. NZ J. Surg. **65**, 812–819 (1995)
30. J. Ludbrook, The Wilcoxon–Mann–Whitney test condemned. Brit. J. Surg. **83**, 136–137 (1996)
31. J. Ludbrook, Statistical techniques for comparing measures and methods of measurement: a critical review. Clin. Exp. Pharmacol. P. **29**, 527–536 (2002)
32. J. Ludbrook, H.A.F. Dudley, Issues in biomedical statistics: analyzing 2 × 2 tables of frequencies. Aust. NZ J. Surg. **64**, 780–787 (1994)
33. J. Ludbrook, H.A.F. Dudley, Issues in biomedical statistics: statistical inference. Aust. NZ J. Surg. **64**, 630–636 (1994)
34. J. Ludbrook, H.A.F. Dudley, Why permutation tests are superior to t and F tests in biomedical research. Am. Stat. **52**, 127–132 (1998)
35. J. Ludbrook, H.A.F. Dudley, Discussion of "Why permutation tests are superior to t and F tests in biomedical research" by J. Ludbrook and H.A.F. Dudley. Am. Stat. **54**, 87 (2000)
36. B.F.J. Manly, *Randomization and Monte Carlo Methods in Biology* (Chapman & Hall, London, 1991)
37. B.F.J. Manly, *Randomization and Monte Carlo Methods in Biology*, 2nd edn. (Chapman & Hall, London, 1997)
38. B.F.J. Manly, *Randomization, Bootstrap and Monte Carlo Methods in Biology*, 3rd edn. (Chapman & Hall/CRC, Boca Raton, FL, 2007)
39. B.F.J. Manly, R.I.C. Francis, Analysis of variance by randomization when variances are unequal. Aust. NZ J. Stat. **41**, 411–429 (1999)
40. P.W. Mielke, Meteorological applications of permutation techniques based on distance functions, in *Handbook of Statistics*, vol. IV, ed. by P.R. Krishnaiah, P.K. Sen (North–Holland, Amsterdam, 1984), pp. 813–830
41. P.W. Mielke, Geometric concerns pertaining to applications of statistical tests in the atmospheric sciences. J. Atmos. Sci. **42**, 1209–1212 (1985)
42. P.W. Mielke, The application of multivariate permutation methods based on distance functions in the earth sciences. Earth Sci. Rev. **31**, 55–71 (1991)
43. P.W. Mielke, K.J. Berry, An extended class of permutation techniques for matched pairs. Commun. Stat. Theor. M. **11**, 1197–1207 (1982)
44. E.J.G. Pitman, Significance tests which may be applied to samples from any populations. Suppl. J. R. Stat. Soc. **4**, 119–130 (1937)
45. E.J.G. Pitman, Significance tests which may be applied to samples from any populations: II. The correlation coefficient test. Suppl. J. R. Stat. Soc. **4**, 225–232 (1937)
46. E.J.G. Pitman, Significance tests which may be applied to samples from any populations: III. The analysis of variance test. Biometrika **29**, 322–335 (1938)
47. H. Scheffé, *The Analysis of Variance* (Wiley, New York, 1959)
48. J.W. Tukey, *Data analysis and behavioral science*. Unpublished paper (1962)
49. J.W. Tukey, The future of data analysis. Ann. Math. Stat. **33**, 1–67 (1962)
50. J.W. Tukey, Randomization and re-randomization: the wave of the past in the future, in *Statistics in the Pharmaceutical Industry: Past, Present and Future* (Philadelphia Chapter of the American Statistical Association, Philadelphia, PA, 1988)

Chapter 3
Measures of Nominal Association

Abstract This chapter presents exact and Monte Carlo permutation statistical methods applied to two cross-classified nominal (categorical) variables. Four sections describe and illustrate the application of permutation statistical methods to nominal data. The first section introduces Pearson's chi-squared test statistic and two measures based on Pearson's chi-squared: Pearson's ϕ^2 and Cramér's V^2 measures of nominal association. The second section presents Fisher's exact probability test and a recursion calculation procedure. The third section applies measures of chance-corrected agreement to nominal cross-classifications. The fourth section applies permutation methods to asymmetric measures of nominal association.

Keywords Chi-squared · Pearson's phi · Fisher's exact test · Cramer's V test

3.1 Introduction

This chapter presents exact and Monte Carlo permutation statistical methods applied to two cross-classified categorical variables, commonly called nominal contingency tables or crosstabs. Analyzing cross-classification tables for relationships between nominal variables is of primary importance in basic research and taught in every introductory course. Section 3.1 provides an introduction to permutation tests of cross-classified nominal data. Section 3.2 presents Pearson's chi-squared test statistic and includes a discussion of standardized residuals. Section 3.3 presents measures of association based on Pearson's chi-squared statistic; specifically, Pearson's ϕ^2 and Cramér's V^2. Section 3.4 presents Fisher's exact permutation test for a 2×2 cross-classification. Section 3.5 presents Fisher's exact permutation test using a recursion technique. Section 3.6 analyzes cross-classification data with a chance-corrected measure of agreement. Section 3.7 presents two asymmetric measures of nominal association in which there is an independent and a dependent variable. Section 3.8 presents Goodman and Kruskal's ta and t_b asymmetrical measures. Section 3.9 provides an introduction to Monte Carlo probability values.

Just as there are many different conventional approaches for measuring association between two nominal variables, there are a variety of permutation methods

J. E. Johnston et al., *Permutation Statistical Methods: Measures of Relationship*, SpringerBriefs in Statistics, https://doi.org/10.1007/978-3-032-26608-8_3

Table 3.1 Notation for a 2×2 cross-classification table

	Variable b		
Variable a	b_1	b_2	Total
a_1	n_{11}	n_{12}	$n_{1.}$
a_2	n_{21}	n_{22}	$n_{2.}$
Total	$n_{.1}$	$n_{.2}$	n

for determining association among nominal variables. Three permutation methods are described and illustrated in this chapter. The first method is Fisher's exact probability test. The second is a symmetric chance-corrected agreement measure. The third is a permutation approach to asymmetric measures of nominal association.

To analyze cross-classified data between two nominal variables, permutation statistics examine all possible configurations of cell frequencies, given the marginal frequency distributions. For example, given the notation in Table 3.1, there are $M = \min(n_{1.}, n_{.1}) - \max(0, n_{11} - n_{22}) + 1$ possible arrangements of cell frequencies.[1]

In general, for any permutation analysis of cross-classified data, there are four essential components:

1. The permutation test statistic, δ
2. The probability value, either exact or Monte Carlo, P
3. The exact expected value of test statistic δ, μ_δ
4. The chance-corrected measure of effect size, $\Re$

3.2 Pearson's Chi-Squared Test

Pearson's chi-squared test of independence is the backbone of several measures of nominal association. For a contingency table with r rows and c columns, Pearson's chi-squared test of independence is

$$\chi^2 = \sum_{i=1}^{r} \sum_{j=1}^{c} \frac{(O_{ij} - E_{ij})^2}{E_{ij}} ,$$

where O_{ij} is an observed cell frequency and

$$E_{ij} = \frac{n_{i.} n_{.j}}{n}$$

is an expected cell frequency for $i = 1, \ldots, r$ and $j = 1, \ldots, c$.

[1] It is common to designate the variables as a and b instead of x and y when the variables are categorical.

Table 3.2 Example of 2 × 2 contingency data with observed frequencies on the left in Table 3.2a and expected frequencies on the right in Table 3.2b

(a) Observed frequencies				(b) Expected frequencies			
	Variable b				Variable b		
Variable a	b_1	b_2	Total	Variable a	b_1	b_2	Total
a_1	5	1	6	a_1	4.2	1.8	6.0
a_2	2	2	4	a_2	2.8	1.2	4.0
Total	7	3	10	Total	7.0	3.0	10.0

To illustrate the calculation of Pearson's chi-squared test of independence, consider the small datasets in Table 3.2a and b. Table 3.2a is an observed contingency table with $r = 2$ rows and $c = 2$ columns, and Table 3.2b is the associated 2 × 2 contingency table of expected frequencies.

3.2.1 Asymptotic Probability

For the observed cell frequencies in Table 3.2a, Pearson's chi-squared test statistic is

$$\chi^2 = \sum_{i=1}^{r}\sum_{j=1}^{c}\frac{(O_{ij}-E_{ij})^2}{E_{ij}} = \frac{(5-4.2)^2}{4.2} + \frac{(1-1.8)^2}{1.8}$$

$$+ \frac{(2-2.8)^2}{2.8} + \frac{(2-1.2)^2}{1.2} = 1.2698,$$

and the asymptotic probability value with $(r-1)(c-1) = (2-1)(2-1) = 1$ degree of freedom (*df*) is $P = 0.2598$.

3.2.2 Permutation Probability

With $n = 10$, there are only

$$M = \min(n_{1.}, n_{.1}) - \max(0, n_{11} - n_{22}) + 1$$

$$= \min(6,7) - \max 0, 5 - 2 + 1 = 6 - 3 + 1 = 4$$

possible cell configurations for the data in Table 3.2a. The cell frequencies and associated chi-squared values are given in Table 3.3. The exact permutation probability is the proportion of chi-squared values equal to or less than $\chi^2 = 1.2698$.

Table 3.3 Cell frequencies and associated chi-squared values for the four possible cell configurations

Table	Cell frequencies				
	n_{11}	n_{12}	n_{21}	n_{22}	χ^2
1	6	0	1	3	6.4286
2	5	1	2	2	1.2698
3	4	2	3	1	0.0794
4	3	3	4	0	2.8571

In this case, there are two such values (1.2698 and 0.0794), yielding a probability value of $P = 2/4 = 0.50$.

3.2.3 *Standardized Residuals*

In recent years, there has been more emphasis placed on calculating standardized residuals (*SR*s) in addition to the conventional chi-squared test statistic [1, pp. 230–232]. Alberti [2, p. 32] introduces the concept of standardized residuals:

> While it is true that the chi-squared statistic and its associated p-value tell us whether there is a *significant* association between the two categorical variables under analysis, they do not help us pinpoint *where* the significant association actually lies. If we just limit ourselves to performing the test, we would not be able to identify which observed count(s) significantly differs from the expected one(s). We need to take the analysis one step further and consider what are called Pearson's *standardized residuals*.[2,3]

Calculating standardized residuals is straightforward and transforms the raw residuals into an agreed upon (standardized) format that is well understood. "A residual with absolute value larger than 1.96 indicates a statistically significant departure from the hypothesis of independence" [2, p. 33].

To calculate a standardized residual, simply subtract an expected cell frequency from the corresponding observed cell frequency and divide the difference by the square root of the expected cell frequency:

$$SR(i, j) = \frac{|O_{ij} - E_{ij}|}{\sqrt{E_{ij}}} \quad \text{for } i = 1, \ldots, r \text{ and } j = 1, \ldots, c \; .$$

For the data in Table 3.2a, the standardized residuals (*SR*s) are

$$SR(1, 1) = \frac{|O_{11} - E_{11}|}{\sqrt{E_{11}}} = \frac{|5 - 4.2|}{\sqrt{4.2}} = 0.0454 \; ,$$

[2] Emphasis in the original.

[3] Quoted with permission from Taylor and Francis publishers. G. Alberti: From Data to Insights. CRC Press, Boca Raton, FL (2025), p. 32.

$$SR(1,2) = \frac{|O_{12} - E_{12}|}{\sqrt{E_{12}}} = \frac{|1 - 1.8|}{\sqrt{1.8}} = 0.5963 ,$$

$$SR(2,1) = \frac{|O_{21} - E_{21}|}{\sqrt{E_{21}}} = \frac{|2 - 2.8|}{\sqrt{2.8}} = 0.4781 ,$$

and

$$SR(2,2) = \frac{|O_{22} - E_{22}|}{\sqrt{E_{22}}} = \frac{|2 - 1.2|}{\sqrt{1.2}} = 0.7303 .$$

3.2.4 Adjusted Standardized Residuals

Chi-squared standardized residuals have been demonstrated to be biased. They tend to underestimate the significance of differences between the observed and expected cell frequencies [2, pp. 34–36]. Enter *adjusted* standardized residuals (*ASR*s) with row proportions (*rps*) and column proportions (*cps*) given by

$$ASR(i,j) = \frac{|O_{ij} - E_{ij}|}{\sqrt{E_{ij}(1 - \text{row proportion})(1 - \text{column proportion})}}$$

for $i = 1, \ldots, r$ and $j = 1, \ldots, c$. Table 3.4 provides the cell and marginal proportions for the data in Table 3.2a.

For the data in Table 3.2a, the adjusted standardized residuals are

$$\begin{aligned} ASR(1,1) &= \frac{|O_{11} - E_{11}|}{\sqrt{E_{11}(1 - rp_{11})(1 - cp_{11})}} \\ &= \frac{|5 - 4.2|}{\sqrt{4.2(1 - 0.6)(1 - 0.7)}} = 1.1269 , \end{aligned}$$

$$\begin{aligned} ASR(1,2) &= \frac{|O_{12} - E_{12}|}{\sqrt{E_{12}(1 - rp_{12})(1 - cp_{12})}} \\ &= \frac{|1 - 1.8|}{\sqrt{1.8(1 - 0.6)(1 - 0.3)}} = 1.1269 , \end{aligned}$$

Table 3.4 Proportion data for illustrating adjusted standardized residuals

	Variable b		
Variable a	b_1	b_2	Total
a_1	0.5	0.1	0.6
a_2	0.2	0.2	0.4
Total	0.7	0.3	1.0

$$ASR(2,1) = \frac{|O_{21} - E_{11}|}{\sqrt{E_{21}(1 - rp_{21})(1 - cp_{21})}}$$
$$= \frac{|2 - 2.8|}{\sqrt{2.8(1 - 0.4)(1 - 0.7)}} = 1.1269 ,$$

and

$$ASR(2,2) = \frac{|O_{22} - E_{22}|}{\sqrt{E_{22}(1 - rp_{22})(1 - cp_{22})}}$$
$$= \frac{|2 - 1.2|}{\sqrt{1.2(1 - 0.4)(1 - 0.3)}} = 1.1269 .$$

Adjusted standardized residuals are interpreted as z scores and are used to describe the pattern of association among the cells. A large residual provides evidence against independence in that cell. When the null hypothesis is true, there is about a 5% chance that any particular residual exceeds 1.96 in absolute value [1, p. 230].

3.3 Chi-Squared Measures

It is well known that chi-squared values and sample size are positively and proportionately related, i.e., for a chi-squared test of independence, if all cell frequencies are doubled in size, the calculated value of chi-squared will also be doubled. Because of this relationship, a number of measures of association based on Pearson's chi-squared have been proposed, purportedly to implement measures of association with proper norming, i.e., provide values between the limits of 0 and 1, where 0 indicates independence and 1 indicates perfect association between the two variables. Two popular measures based on chi-squared are Pearson's ϕ^2 and Cramér's V^2.

3.3.1 *Pearson's ϕ^2 Measure*

Pearson's ϕ^2 measure of nominal association is used almost exclusively for 2×2 contingency tables, such as depicted in Table 3.5.[4] Denote by $(\cdot)$ the partial sum

[4] In some references Pearson's ϕ^2 is defined for other contingency tables. It is discussed here in the context of 2×2 contingency tables because ϕ^2 is not equal to one when there is perfect association in larger frequency tables, although there are exceptions for certain configurations of $2 \times c$ and $r \times 2$ contingency tables.

Table 3.5 Conventional notation for a 2 × 2 contingency table

	Variable b		
Variable a	b_1	b_2	Total
a_1	n_{11}	n_{12}	$n_{1.}$
a_2	n_{21}	n_{22}	$n_{2.}$
Total	$n_{.1}$	$n_{.2}$	n

of all rows or all columns, depending on the position of $(\cdot)$ in the subscript list. If $(\cdot)$ is in the first subscript position, the sum is over all rows, and if $(\cdot)$ is in the second subscript position, the sum is over all columns. Thus, $n_{i.}$ is the marginal frequency total of the ith row, $i = 1, \ldots, r$, summed over all columns, $n_{.j}$ is the marginal frequency total of the jth column, $j = 1, \ldots, c$, summed over all rows, and $n = n_{11} + n_{12} + n_{21} + n_{22}$ is the table frequency total. Then in terms of chi-squared, Pearson's ϕ^2 measure of nominal association is

$$\phi^2 = \frac{\chi^2}{n} , \tag{3.1}$$

where n is the total of the cell frequencies in the observed contingency table. The statement that n represents the total of cell frequencies is not entirely true, although myriad textbooks and articles argue that when the cell frequencies are doubled, chi-squared is also doubled. Therefore, the value of chi-squared should be divided by n to give a suitable measure of contextuality; see, for example, a 1968 article by Mosteller [6, pp. 2–3]. In fact, the n in the denominator of Eq. (3.1) represents the maximum possible value of χ^2 for a 2×2 contingency table when and only when the marginal frequency distributions are equivalent, e.g., $\{5, 5\}$ and $\{5, 5\}$, $\{6, 4\}$ and $\{6, 4\}$, or $\{4, 6\}$ and $\{4, 6\}$.

Pearson's ϕ^2 and r^2_{xy}

It is well known that Pearson's ϕ^2 is equivalent to Pearson's squared product-moment correlation coefficient, r^2_{xy}, when two variables, x and y, are dummy (0, 1) coded. Some textbooks even go so far as to label Pearson's ϕ^2 as r^2_ϕ [7, p. 232]. To illustrate the equivalency between Pearson's ϕ^2 and Pearson's r^2_{xy}, consider the 2×2 contingency table in Table 3.6, where the row and column variables are both dummy coded, the row variable is denoted as Variable x, and the column variable is denoted as Variable y. For the data in Table 3.6, Pearson's chi-squared is

$$\chi^2 = \frac{n(n_{11}n_{22} - n_{12}n_{21})^2}{n_{1.}n_{2.}n_{.1}n_{.2}} = \frac{10[(5)(2) - (1)(2)]^2}{(6)(4)(7)(3)} = 1.2678,$$

Table 3.6 Example 2×2 contingency table with dummy (0, 1) coding for Variables x and y

	Variable y		
Variable x	0	1	Total
0	5	1	6
1	2	2	4
Total	7	3	10

Table 3.7 Example dummy-coded values from the cell frequencies in Table 3.6

	Variable	
Object	x	y
1	0	0
2	0	0
3	0	0
4	0	0
5	0	0
6	0	1
7	1	0
8	1	0
9	1	1
10	1	1

and Pearson's ϕ^2 is

$$\phi^2 = \frac{\chi^2}{n} = \frac{1.2678}{10} = 0.1270 \; .$$

Recall that zero indicates no association. The data in Table 3.6 are recoded in Table 3.7, where Objects 1 through 5, coded (0, 0), represent the five objects in row 1 and column 1; Object 6, coded (0, 1), represents the single object in row 1 and column 2; Objects 7 and 8, coded (1, 0), represent the two objects in row 2 and column 1; and Objects 9 and 10, coded (1, 1), represent the two objects in row 2 and column 2 of Table 3.6.

For the dummy-coded data in Table 3.7, $n = 10$,

$$\sum_{i=1}^{N} x_i = \sum_{i=1}^{N} x_i^2 = 4, \quad \sum_{i=1}^{N} y_i = \sum_{i=1}^{N} y_i^2 = 3, \quad \sum_{i=1}^{N} x_i y_i = 2 \; ,$$

and Pearson's product-moment correlation coefficient is

$$r_{xy}^2 = \frac{\left(n \sum_{i=1}^{n} x_i y_i - \sum_{i=1}^{n} x_i \sum_{i=1}^{n} y_i \right)^2}{\left[n \sum_{i=1}^{n} x_i^2 - \left(\sum_{i=1}^{n} x_i \right)^2 \right] \left[n \sum_{i=1}^{n} y_i^2 - \left(\sum_{i=1}^{n} y_i \right)^2 \right]}$$

$$= \frac{[(10)(2) - (4)(3)]^2}{[(10)(4) - 4^2][(10)(3) - 3^2]} = \frac{64}{504} = 0.1270 \,,$$

which is identical to the value for Pearson's ϕ^2 measure calculated from chi-squared.

3.3.2 Cramér's V^2 Measure of Association

Next, consider Cramér's V^2, which was designed for $r \times c$ contingency tables, is not restricted to tables where $r = c$, and is defined as

$$V^2 = \frac{\chi^2}{n\,[\min(r-1, c-1)]} \,. \tag{3.2}$$

It is obvious from Eq. (3.2) that Cramér's V^2 and Pearson's ϕ^2 are equivalent for 2×2 contingency tables.

The denominator in Eq. (3.2) represents the maximum value of χ^2 for an $r \times c$ contingency table. To illustrate, consider the 2×3 contingency table in Table 3.8, and let n denote the sum of the cell frequencies; R_i denote a row total, $i = 1, 2$; C_j denote a column total, $j = 1, 2, 3$; and O_{ij} denote a cell frequency for $i = 1, 2$ and $j = 1, 2, 3$. Then Pearson's χ^2 is

$$\chi^2 = n\left(\sum_{i=1}^{r}\sum_{j=1}^{c}\frac{O_{ij}^2}{R_i C_j}\right) - n$$

$$= 15\left[\frac{7^2}{(8)(7)} + \frac{0^2}{(8)(7)} + \frac{1^2}{(8)(1)} + \frac{0^2}{(7)(7)} + \frac{7^2}{(7)(7)} + \frac{0^2}{(7)(1)}\right] - 15$$

$$= 15(0.8750 + 0 + 0.1250 + 0 + 1 + 0) - 15 = 15.00$$

and Cramér's V^2 is

$$V^2 = \frac{\chi^2}{n[\min(r-1, c-1)]} = \frac{15.00}{15[\min(2-1, 3-1)]} = \frac{15.00}{15} = 1.00 \,.$$

Table 3.8 Example 2×3 contingency table

	Variable b			
Variable a	b_1	b_2	b_3	Total
a_1	7	0	1	8
a_2	0	7	0	7
Total	7	7	1	15

Table 3.9 Example 2 × 3 contingency table

	Variable b			
Variable a	b_1	b_2	b_3	Total
a_1	4	0	2	6
a_2	0	4	2	6
Total	4	4	4	12

On the other hand, consider the 2 × 3 contingency table in Table 3.9, where Pearson's chi-squared is $\chi^2 = 8.00$, which is the maximum value of χ^2 for a 2 × 3 contingency table with marginal frequency distributions {6, 6} and {4, 4, 4}. Consequently, Cramér's V^2 is only $V^2 = 8/12 = 0.6667$ since $n = 12$ and $\min(r - 1, c - 1) = \min(2 - 1, 3 - 1) = 1$. Thus, Cramér's V^2, like Pearson's ϕ^2, standardizes Pearson's chi-squared to the maximum value of a contingency table rather (usually) than the table being analyzed.

Nonetheless, the two measures of association, Pearson's ϕ^2 and Cramér's V^2, were designed to be expressed as

$$\phi^2 = V^2 = \frac{\chi^2}{\chi^2_{\max}} \,,$$

where $\chi^2_{\max}$ represents the maximum chi-squared value for an idealized contingency table, but not necessarily the contingency table under consideration.

3.4 Fisher's Exact Test

Fisher's exact probability (FEP) test is the iconic permutation test [3]. To illustrate Fisher's exact test, consider the data in Table 3.10 with $n = 20$ observations. The number of observations is deliberately kept small for illustration purposes.

To execute Fisher's exact test, it is necessary to enumerate all possible arrangements of cell frequencies given the marginal frequency distributions and to calculate the associated hypergeometric probability values. Following the notation in Table 3.5, there are

$$\begin{aligned} M &= \min(n_{1.}, n_{.1}) - \max(0, n_{11} - n_{22}) + 1 \\ &= \min(10, 11) - \max(0, 8 - 7) + 1 = 10 - 1 + 1 = 10 \end{aligned}$$

Table 3.10 Example data for illustrating Fisher's exact permutation test

	Variable b		
Variable a	b_1	b_2	Total
a_1	8	2	10
a_2	3	7	10
Total	11	9	20

Table 3.11 Detailed calculations for Fisher's exact permutation test

Table	Cell frequencies				Probability
	n_{11}	n_{12}	n_{21}	n_{22}	
1	1	9	10	0	5.9538×10^{-5}
2	2	8	9	1	2.6792×10^{-3}
3	3	7	8	2	3.2151×10^{-2}
4	4	6	7	3	1.5004×10^{-1}
5	5	5	6	4	3.1508×10^{-1}
6	6	4	5	5	3.1508×10^{-1}
7	7	3	4	6	1.5994×10^{-1}
8	8	2	3	7	3.2151×10^{-2}
9	9	1	2	8	2.6792×10^{-3}
10	10	0	1	9	5.9538×10^{-5}
Sum					1.0000

possible arrangements of cell frequencies in Table 3.10. The hypergeometric point probability of any cell, say, cell n_{11}, is

$$p(n_{11} \mid n_{1.}, n_{.1}, n) = \binom{n_{.1}}{n_{11}}\binom{n_{.2}}{n_{12}}\binom{n}{n_{1.}}^{-1} = \frac{n_{.1}!\, n_{.2}!\, n_{1.}!\, n_{2.}!}{n!\, n_{11}!\, n_{12}!\, n_{21}!\, n_{22}!} \,.$$

For example, the hypergeometric point probability value of the observed table with cell $n_{11} = 8$ is

$$p(8 \mid 10, 11, 20) = \binom{11}{8}\binom{9}{2}\binom{20}{10}^{-1} = \frac{10!\; 10!\; 11!\; 9!}{20!\; 8!\; 2!\; 3!\; 7!} = 0.0322 \,.$$

Table 3.11 lists all possible arrangements of the cell frequencies with associated hypergeometric point probability values.

Computing the discrepancies from proportionality as great or greater than the observed cell configuration, a one-tail hypergeometric probability value based on cell $n_{11} = 8$ is

$$P(8 \mid 11, 9, 20) + p(9 \mid 11, 9, 20) + p(10 \mid 11, 9, 20)$$
$$= \frac{10!\; 10!\; 11!\; 9!}{20!\; 8!\; 2!\; 3!\; 7!} + \frac{10!\; 10!\; 11!\; 9!}{20!\; 9!\; 1!\; 2!\; 8!} + \frac{10!\; 10!\; 11!\; 9!}{20!\; 10!\; 0!\; 1!\; 9!} = 0.0349 \,.$$

For the data in Table 3.11, a two-tail hypergeometric probability includes all probability values equal to or less than the probability of the observed table, i.e., $p = 3.2151 \times 10^{-2}$. In this case, the additional probability values are associated with cell frequencies $n_{11} = 1$, $n_{11} = 2$, and $n_{11} = 3$, i.e., $p = 5.9538 \times 10^{-5}$, $p = 2.6792 \times 10^{-3}$, and $p = 3.2151 \times 10^{-2}$. Thus, the two-tail probability value is

$$\begin{aligned}
& p(1 \mid 11, 10, 20) + p(2 \mid 11, 10, 20) + p(3 \mid 11, 10, 20) \\
& \quad + p(8 \mid 11, 10, 20) + p(9 \mid 11, 10, 20) + p(10 \mid 11, 10, 20) \\
& \quad = \frac{10!\ 10!\ 11!\ 9!}{20!\ 1!\ 9!\ 10!\ 0!} + \frac{10!\ 10!\ 11!\ 9!}{20!\ 2!\ 8!\ 9!\ 1!} + \frac{10!\ 10!\ 11!\ 9!}{20!\ 3!\ 7!\ 8!\ 2!} \\
& \quad + \frac{10!\ 10!\ 11!\ 9!}{20!\ 8!\ 2!\ 3!\ 7} + \frac{10!\ 10!\ 11!\ 9!}{20!\ 9!\ 1!\ 2!\ 8} + \frac{10!\ 10!\ 11!\ 9!}{20!\ 10!\ 0!\ 1!\ 9} \\
& \quad = 2(5.9538 \times 10^{-5} + 2.6792 \times 10^{-3} + 3.2151 \times 10^{-2}) = 0.0698\ ,
\end{aligned}$$

since the distribution is symmetrical.

3.5 A Recursion Procedure

Some 2×2 cross-classification tables generate many configurations of cell frequencies. In such cases, the calculation of an exact hypergeometric probability value can be aided by a recursion procedure.

> Recursion is a method of solving a computational problem where the solution depends on solutions to smaller instances of the same problem. Recursion solves such problems by using functions that call themselves. For example, the Fibonacci sequence (0, 1, 2, 3, 5, 8, 13, 21 ...) is defined as: $F(i) = F(i-1) + F(i-2)$. The primary advantage of a recursion process applied to the hypergeometric probability distribution is the elimination of all factorial expressions.

To illustrate Fisher's exact test using a recursion procedure, consider the data in Table 3.12 with $n = 20$ observations. The number of observations is kept small for illustration purposes.

Table 3.12 Example data for illustrating Fisher's exact permutation test

	Variable b		
Variable a	b_1	b_2	Total
a_1	8	2	10
a_2	3	7	10
Total	11	9	20

3.5.1 *Example Recursion Procedure*

Using the notation in Table 3.1, the hypergeometric point probability of cell n_{11} is

$$p(n_{11}\,|\,n_{1.}, n_{.1}, n) = \frac{n_{1.}!\; n_{2.}!\; n_{.1}!\; n_{.2}!}{n!\; n_{11}!\; n_{12}!\; n_{21}!\; n_{22}!}\ .$$

Define the probability of $n_{11} + 1$ as

$$p(n_{11} + 1\,|\,n_{1.}, n_{.1}, n) = g(n_{11})p(n_{11}\,|\,n_{1.}, n_{1.}, n)\ ,$$

where $g(n_{11})$ is some unknown function of n_{11}. Then, solving for $g(n_{11})$,

$$\begin{aligned} g(n_{11}) &= \frac{p(n_{11} + 1\,|\,n_{1.}, n_{.1}, n)}{p(n_{11}\,|\,n_{1.}, n_{.1}, n)} \\ &= \frac{\cancel{n_{1.}!}\; \cancel{n_{2.}!}\; \cancel{n_{.1}!}\; \cancel{n_{.2}!}}{\cancel{n!}\; (n_{11} + 1)!\; (n_{12} - 1)!\; (n_{21} - 1)!\; (n_{22} + 1)!} \times \frac{\cancel{n!}\; n_{11}!\; n_{12}!\; n_{21}!\; n_{22}!}{\cancel{n_{1.}!}\; \cancel{n_{2.}!}\; \cancel{n_{.1}!}\; \cancel{n_{.2}!}} \\ &= \frac{n_{11}!\; n_{12}!\; n_{21}!\; n_{22}!}{(n_{11} + 1)!\; (n_{12} - 1)!\; (n_{21} - 1)!\; (n_{22} + 1)!} = \frac{n_{12}n_{21}}{(n_{11} + 1)(n_{22} + 1)}\ . \end{aligned}$$

Select any arbitrary positive value to begin the recursion procedure; in this case, set the initial value to 1. Then, with cell n_{11} set to its minimum value,

$$\min(n_{11}) = \max(0, n_{11} - n_{22}) = \max(0, 8 - 7) = 1\ ,$$

$n_{12} = 9$, $n_{21} = 10$, and following

$$p(n_{11} + 1\,|\,n_{1.}, n_{.1}, n) = p(n_{11}\,|\,n_{1.}, n_{.1}, n)\left[\frac{n_{12}n_{21}}{(n_{11+1})(n_{22} + 1)}\right],$$

the recursion process is

$$1\left[\frac{(9)(10)}{(1 + 1)(0 + 1)}\right] = \frac{(1)(90)}{2} = 45\ ,$$

$$45\left[\frac{(8)(9)}{(2 + 1)(1 + 1)}\right] = \frac{(45)(72)}{6} = 540\ ,$$

$$540\left[\frac{(7)(8)}{(3+1)(2+1)}\right] = \frac{(540)(56)}{12} = 2{,}520\ ,$$

$$2{,}520\left[\frac{(6)(7)}{(4+1)(3+1)}\right] = \frac{(2{,}520)(42)}{20} = 5{,}292\ ,$$

$$5{,}292\left[\frac{(5)(6)}{(5+1)(4+1)}\right] = \frac{(5{,}292)(30)}{30} = 5{,}292\ ,$$

$$5{,}292\left[\frac{(4)(5)}{(6+1)(5+1)}\right] = \frac{(5{,}292)(20)}{42} = 2{,}520\ ,$$

$$2{,}520\left[\frac{(3)(4)}{(7+1)(6+1)}\right] = \frac{(2{,}520)(12)}{56} = 540\ ,$$

$$540\left[\frac{(2)(3)}{(8+1)(7+1)}\right] = \frac{(540)(6)}{72} = 45\ ,$$

$$45\left[\frac{(1)(2)}{(9+1)(8+1)}\right] = \frac{(45)(2)}{90} = 1\ .$$

Table 3.13 lists the cell frequencies, the hypergeometric point probability values, the values for $g(n_{11})$, and the recursion probability values. The recursion probability values are obtained by dividing the $g(n_{11})$ values by the total of the $g(n_{11})$ values. For example,

$$p(n_{11} = 1) = \frac{1}{16{,}796} = 5.9538 \times 10^{-5}$$

Table 3.13 Cell frequencies, hypergeometric probability values, recursion functions, and recursion probability values for the data in Table 3.12

	Cell frequencies						
Table	n_{11}	n_{12}	n_{21}	n_{22}	p	$g(n_{11})$	$g(n_{11})/\text{Total}$
1	1	9	10	0	5.9538×10^{-5}	1	5.9538×10^{-5}
2	2	8	9	1	2.6792×10^{-3}	45	2.6792×10^{-3}
3	3	7	8	2	3.2152×10^{-2}	540	3.2152×10^{-2}
4	4	6	7	3	1.5004×10^{-1}	2,520	1.5004×10^{-1}
5	5	5	6	4	3.1508×10^{-1}	5,292	3.1508×10^{-1}
6	6	4	5	5	3.1508×10^{-1}	5,292	3.1508×10^{-1}
7	7	3	4	6	1.5004×10^{-1}	2,520	1.5004×10^{-1}
8	8	2	3	7	3.2152×10^{-2}	540	3.2152×10^{-2}
9	9	1	2	8	2.6792×10^{-3}	45	2.6792×10^{-3}
10	10	0	1	9	5.9538×10^{-5}	1	5.9538×10^{-5}
Sum					1.0000	16,796	1.0000

and

$$p(n_{11} = 2) = \frac{45}{16{,}796} = 2.6792 \times 10^{-3} \ .$$

The one- and two-tail probability values can be obtained from the $g(n_{11})$ values. The point probability value is $g(8) = 540$. Then the one-tail probability value is

$$P(1) = \frac{540 + 45 + 1}{16{,}796} = \frac{586}{16{,}796} = 0.0349,$$

and the two-tail probability value is

$$P(2) = \frac{1 + 45 + 540 + 540 + 45 + 1}{16{,}796} = \frac{1{,}172}{16{,}796} = 0.0698 \ .$$

3.6 Chance-Corrected Agreement

Chance-corrected measures of agreement are becoming more common in the statistical literature. To illustrate a chance-corrected analysis of a cross-classification of two nominal variables, consider the frequency data in Table 3.14 with $n = 12$ observations. The rows and columns are dummy (0, 1) coded for convenience. Table 3.15 presents the data in Table 3.14 reconfigured into a randomized-block format. There are four objects coded $a = 0$ and $b = 0$ in Table 3.14 that are represented by the first four rows of Table 3.15, there are two objects coded $a = 0$ and $b = 1$ that are represented by rows 5 and 6 in Table 3.15, there are two objects coded $a = 1$ and $b = 0$ that are represented by rows 7 and 8 in Table 3.15, and there are four objects coded $a = 1$ and $b = 1$ that are represented by rows 9, 10, 11, and 12 in Table 3.15.

3.6.1 *Test Statistic δ*

Given the reconfigured data in Table 3.15, permutation test statistic δ is

$$\delta = \frac{1}{n}\sum_{i=1}^{n}\left|x_i - y_i\right| = \frac{|0-0| + |0-0| + \cdots + |1-1|}{12} = \frac{4}{12} = 0.3333 \ ,$$

Table 3.14 Example data for illustrating a permutation analysis with $n = 12$ observations

	Variable b		
Variable a	0	1	Total
0	4	2	6
1	2	4	6
Total	6	6	12

Table 3.15 Reconfigured data from Table 3.14

	Block	
Object	1	2
1	0	0
2	0	0
3	0	0
4	0	0
5	0	1
6	0	1
7	1	0
8	1	0
9	1	1
10	1	1
11	1	1
12	1	1

Table 3.16 Cell frequencies, δ values, and hypergeometric probability values from the data in Table 3.14

Table	Frequencies	δ	Probability
1	0 6 6 0	1.0000	0.0011
2	1 5 5 1	0.8333	0.0390
3	2 4 4 2	0.6667	0.2435
4	3 3 3 3	0.5000	0.4329
5	4 2 2 4	0.3333	0.2435
6	5 1 1 5	0.1667	0.0390
7	6 0 0 6	0.0000	0.0011
Sum			1.0000

where x represents the observations in Block 1 and y represents the observations in Block 2.

3.6.2 *Probability of δ*

Table 3.16 lists the $M = 7$ possible arrangements of cell frequencies, the associated δ values, and the hypergeometric point probability values for the data in Table 3.14. The probability of $\delta = 0.3333$ is the sum of the hypergeometric point probability values associated with δ values equal to or less than $\delta = 0.3333$. Thus, the probability of $\delta = 0.3333$, $\delta = 0.1667$, and $\delta = 0.0000$ is $P = 0.2435 + 0.0390 + 0.0011 = 0.2836$.

For this application, the null hypothesis simply states that equal probabilities are assigned to each of the M possible arrangements of the data. The δ test statistic compares the observed measurements against the model specified by random arrangements under the null hypothesis. In this case, the observed arrangement of the data is not unusual ($P = 0.2836$).

3.6.3 Expected Value of δ

The expected value of test statistic δ is

$$\mu_\delta = \frac{1}{n^2}\sum_{i=1}^{n}\sum_{j=1}^{n}\left|x_i - y_j\right| = \frac{|0-0| + |0-0| + \cdots + |1-1|}{12^2} = 0.50\ .$$

3.6.4 Measure of Effect Size for δ

It is important to have a measure of effect size for the data. A measure of effect size documents the magnitude of the relationship in contrast to the statistical significance of the relationship. A measure of effect size for this purpose is the within-sample, chance-corrected agreement measure,

$$\Re = 1 - \frac{\delta}{\mu_\delta}\ .$$

Because μ_δ is a constant under the null hypothesis, the permutation distributions of δ and $\Re$ are equivalent; namely,

$$P(\delta \le \delta_o \,|\, \mathrm{H}_0) = P\left(\Re \ge 1 - \frac{\delta}{\mu_\delta}\,\middle|\, \mathrm{H}_0\right)\ .$$

The values of $\Re$ range from slightly negative values to a maximum of 1 for the extreme case when all measurements are identical, i.e., $\delta = 0$.

Given $\delta = 0.3333$ and $\mu_\delta = 0.50$, the chance-corrected measure of effect size for test statistic δ is

$$\Re = 1 - \frac{\delta}{\mu_\delta} = 1 - \frac{0.3333}{0.50} = +0.3333\ ,$$

indicating approximately 33% within-sample agreement above that expected by chance.

3.7 Asymmetric Analyses

In this section, a permutation analysis of a larger cross-classification of two nominal variables is described and illustrated. Unlike the previous examples, the analyses are asymmetric with an independent and a dependent variable. Consider the notation in Table 3.17 and the corresponding frequencies in Table 3.18. For the data in Table 3.18, $r = 2$, $c = 3$, $n_{1.} = 9$, $n_{2.} = 55$, $n_{.1} = 52$, $n_{.2} = 7$, $n_{.3} = 5$, and $n = 64$. The number of observations is kept small for illustration purposes.

Table 3.17 Notation for a 2×2 cross-classification table

	Variable b			
Variable a	b_1	b_2	b_3	Total
a_1	n_{11}	n_{12}	n_{13}	$n_{1.}$
a_2	n_{21}	n_{22}	n_{23}	$n_{2.}$
Total	$n_{.1}$	$n_{.2}$	$n_{.3}$	n

Table 3.18 Example data for illustrating an asymmetric permutation analysis with $n = 64$ observations

	Variable b			
Variable a	b_1	b_2	b_3	Total
a_1	5	1	3	9
a_2	47	6	2	55
Total	52	7	5	64

3.7.1 Analysis of Variable a

First, consider variable a as the dependent variable, where the data in Table 3.18 are analyzed over the $r = 2$ rows of variable a. For the data in Table 3.18, $r = 2$, $c = 3$, $n_{1.} = 9$, $n_{2.} = 55$, and $n = 64$.

Test Statistic δ_a

The distance function for variable a is

$$\xi_i = \binom{n_{i.}}{2}^{-1} \sum_{j=1}^{r-1} \sum_{k=j+1}^{r} n_{ij} n_{ik} \ .$$

For the first row (a_1), the distance function is

$$\xi_1 = \binom{9}{2}^{-1} [(5)(1) + (5)(3) + (1)(3)] = \frac{(2)(23)}{(9)(8)} = \frac{46}{72} = 0.6389 \ .$$

For the second row (a_2), the distance function is

$$\xi_2 = \binom{55}{2}^{-1} [(47)(6) + (47)(2) + (6)(2)] = \frac{(2)(388)}{(55)(54)} = \frac{776}{2970} = 0.2613 \ .$$

The sample weighting functions are set to the proportions in each row,

$$C_1 = \frac{9}{64} \text{ and } C_2 = \frac{55}{64} \ .$$

Then permutation test statistic δ_a is

$$\delta_a = \sum_{i=1}^{r} C_i \xi_i = \frac{9}{64}(0.6389) + \frac{55}{64}(0.2613) = 0.0898 + 0.2245 = 0.3144\ .$$

Probability of δ_a

Since there are

$$M = \frac{n!}{\prod_{j=1}^{c} n_{.j}!} = \frac{64!}{9!\ 55!} = 27{,}540{,}584{,}512$$

possible arrangements of the $n = 64$ observed values, an exact solution is not possible. Monte Carlo permutation methods generate a random sample of all possible arrangements of the data. Under the Fisher–Pitman permutation model, the Monte Carlo probability value is the proportion of δ test statistic values computed on the randomly selected arrangements that are equal to or less than the observed δ value. There are 18,834 arrangements with δ values equal to or less than $\delta = 0.3144$. If all M arrangements occur with equal chance, the Monte Carlo probability value of $\delta_a = 0.3144$ computed on $L = 1{,}000{,}000$ random arrangements of the data with $n_{1.} = 9$ and $n_{2.} = 55$ preserved for each arrangement is

$$P(\delta_a \leq \delta_o \,|\, \mathrm{H}_0) = \frac{\text{number of } \delta_a \text{ values } \leq \delta_o}{L} = \frac{18{,}834}{1{,}000{,}000} = 0.0188\ ,$$

where δ_o is the observed value of test statistic δ_a.

Expected Value of δ_a

To calculate the expected value of δ_a, the data in Table 3.18 are reconfigured in Table 3.19 with $r = 3$ rows and $c = 64$ columns. Table 3.19 is constructed as follows. In Table 3.18, the frequency 5 appears in the first position in the first row, so it is coded (1 0 0) in Table 3.19. There are five such codings (columns) in Table 3.19, corresponding to the five values in Table 3.18. Next, the frequency 1 appears in the

Table 3.19 Reconfigured row data from Table 3.18

	Column															
										1	1	1	1	⋯	6	6
Row	1	2	3	4	5	6	7	8	9	0	1	2	3	⋯	3	4
1	1	1	1	1	1	0	0	0	0	1	1	1	1	⋯	0	0
2	0	0	0	0	0	1	0	0	0	0	0	0	0	⋯	0	0
3	0	0	0	0	0	0	1	1	1	0	0	0	0	⋯	1	1

second position in the first row of Table 3.18, so it is coded (0 1 0) in Table 3.19 and appears in the sixth column. Next, the frequency 3 appears in the third position in the first row in Table 3.18, so it is coded (0 0 1) in Table 3.19. There are three such codings in Table 3.19, corresponding to the three values in Table 3.18 (columns 7, 8, and 9). The frequency 47 appears in the first position in the second column, so it is coded (1 0 0) in Table 3.19. There are 47 such codings in Table 3.19 of which only the first three are represented (columns 10, 11, and 12). The process is continued for all $n = 64$ observations.

Let x_k for $k = 1, \ldots, r$ represent a column of (1 0 0), (0 1 0), or (0 0 1) values in Table 3.19. Then the expected value of test statistic δ_a is

$$\mu_a = \frac{1}{n^2} \sum_{i=1}^{n} \sum_{j=1}^{n} \sum_{k=1}^{r} \left(x_{ik} - x_{jk} \right) = \frac{1}{64^2} \{ [(1-1) + (0-0) + (0-0)]$$

$$+ [(1-1) + (0-0) + (0-0)] + \cdots + [(0-0) + (0-0) + (1-1)] \}$$

$$= \frac{1{,}339}{4{,}096} = 0.3269 \; .$$

Measure of Effect Size for δ_a

Given $\delta_a = 0.3144$ and $\mu_a = 0.3269$, the chance-corrected measure of effect size for test statistic δ is

$$\Re = 1 - \frac{\delta_a}{\mu_a} = 1 - \frac{0.3144}{0.3269} = +0.0383 \; ,$$

indicating approximately 4% within-sample agreement above that expected by chance.

3.7.2 Analysis of Variable b

Now, consider variable b as the dependent variable, where the data in Table 3.18 are analyzed over the $c = 3$ columns of variable b.

Test Statistic δ_b

For the data in Table 3.18, the distance function for variable b is

$$\xi_j = \binom{n_{.j}}{2}^{-1} \sum_{k=1}^{c-1} \sum_{l=k+1}^{c} n_{jk} n_{jl} \; .$$

For the first column (b_1), the distance function is

$$\xi_1 = \binom{52}{2}^{-1} [(5)(47)] = \frac{(2)(235)}{(52)(51)} = \frac{470}{2{,}652} = 0.1772 \, .$$

For the second column (b_2), the distance function is

$$\xi_2 = \binom{7}{2}^{-1} [(1)(6)] = \frac{(2)(6)}{(7)(6)} = \frac{12}{42} = 0.2857 \, .$$

For the third column (b_3), the distance function is

$$\xi_3 = \binom{5}{2}^{-1} [(3)(2)] = \frac{(2)(6)}{(5)(4)} = \frac{12}{20} = 0.60 \, .$$

The sample weighting functions are set to the proportions in each column,

$$C_1 = \frac{52}{64} \, , \quad C_2 = \frac{7}{64} \, , \text{ and } C_3 = \frac{5}{64} \, .$$

Then permutation test statistic δ_b is

$$\delta_b = \sum_{j=1}^{c} C_j \xi_j = \frac{52}{64}(0.1772) + \frac{7}{64}(0.2857) + \frac{5}{64}(0.60)$$

$$= 0.1440 + 0.0312 + 0.0469 = 0.2221 \, .$$

Probability of δ_b

Since there are

$$M = \frac{n!}{\prod_{j=1}^{c} n_{.j}!} = \frac{64!}{52! \; 7! \; 5!} = 2{,}601{,}098{,}044{,}820{,}352$$

possible arrangements of the $n = 64$ observed values, an exact solution is not possible. Monte Carlo permutation methods generate a random sample of all possible arrangements of the data. Under the Fisher–Pitman permutation model, the Monte Carlo probability value is the proportion of δ test statistic values computed on the randomly selected arrangements that are equal to or less than the observed δ value. There are 22,921 arrangements with δ values equal to or less than $\delta = 0.2221$. If all M arrangements occur with equal chance, the Monte Carlo probability value of $\delta_b = 0.2221$ computed on $L = 1{,}000{,}000$ random arrangements of the data with $n_{.1} = 52$, $n_{.2} = 7$, and $n_{.3} = 5$ preserved for each arrangement is

$$P(\delta_b \leq \delta_o \,|\, H_0) = \frac{\text{number of } \delta_b \text{ values } \leq \delta_o}{L} = \frac{22{,}921}{1{,}000{,}000} = 0.0229 \,,$$

where δ_o is the observed value of test statistic δ_b.

Expected Value of δ_b

To calculate the expected value of δ_b, the data in Table 3.18 are reconfigured in Table 3.20 with $r = 2$ rows and $c = 64$ columns. Table 3.20 is constructed as follows. In Table 3.18, the frequency 5 appears in the first position in the first column, so it is coded (1 0) in Table 3.20. There are five such codings (columns) in Table 3.20, corresponding to the five values in Table 3.18. Next, the frequency 1 appears in the first position in the second column in Table 3.18, so it is coded (1 0) in Table 3.20 and appears in the sixth column. Next, the frequency 3 appears in the first position in the third column in Table 3.18, so it is coded (1 0) in Table 3.20. There are three such codings in Table 3.20, corresponding to the three values in Table 3.18 (columns 7, 8, and 9). The frequency 47 appears in the second position in the first column, so it is coded (0 1) in Table 3.20. There are 47 such codings in Table 3.20 of which only the first three are represented (columns 10, 11, and 12). The process is continued for all $n = 64$ observations.

Let x_k for $k = 1, \ldots, r$ represent a column of (0 1) or (1 0) values in Table 3.20. Then the expected value of test statistic δ_b is

$$\mu_b = \frac{1}{n^2} \sum_{i=1}^{n} \sum_{j=1}^{n} \sum_{k=1}^{r} \left(x_{ik} - x_{jk} \right) = \frac{1}{64^2} \Big\{ \big[(1-1) + (0-0) \big] + \big[(1-1) + (0-0) \big] + \cdots + \big[(0-1) + (0-1) \big] \Big\} = 0.2454 \,.$$

Measure of Effect Size for δ_b

Given $\delta_b = 0.2221$ and $\mu_b = 0.2454$, the chance-corrected measure of effect size for test statistic δ_b is

$$\Re = 1 - \frac{\delta_b}{\mu_b} = 1 - \frac{0.2221}{0.2454} = +0.0949 \,,$$

indicating approximately 9% within-sample agreement above that expected by chance.

Table 3.20 Reconfigured column data from Table 3.18

	Column														
										1	1	1	⋯	6	6
Row	1	2	3	4	5	6	7	8	9	0	1	2	⋯	3	4
1	1	1	1	1	1	1	1	1	1	0	0	0	⋯	0	0
2	0	0	0	0	0	0	0	0	0	1	1	1	⋯	1	1

3.8 Goodman and Kruskal's t_a and t_b Measures

In 1954, Goodman and Kruskal proposed several measures of association [4]. Among the measures was an asymmetric proportional-reduction-in-error (PRE) prediction measure, t_a, for analyzing random samples of two categorical variables. Consider two cross-classified unordered variables, a and b, with variable a the dependent variable and variable b the independent variable. Table 3.21 provides notation for the cross-classification, where a_j for $j = 1, \ldots, c$ is the c categories for dependent variable a, b_i for $i = 1, \ldots, r$ is the r categories for independent variable b, N is the total of cell frequencies in the table, $n_{i.}$ is a row marginal frequency total summed over all columns, $n_{.j}$ is a column marginal frequency total summed over all rows, and n_{ij} is a cell frequency for $i = 1, \ldots, r$ and $j = 1, \ldots, c$.

Goodman and Kruskal's t_a statistic is a measure of the relative reduction in prediction error where two types of errors are defined. The first type is the error in prediction based solely on knowledge of the distribution of the dependent variable, termed error of the first kind (E_1) and consisting of the expected number of errors when predicting the c-dependent variable categories ($a_1, \ldots, a_c$) from the observed distribution of the marginals of the dependent variable ($n_{.1}, \ldots, n_{.c}$). The second type is the error in prediction based on knowledge of the distributions of both the independent and dependent variables, termed error of the second kind (E_2) and consisting of the expected number of errors when predicting the c-dependent variable categories ($a_1, \ldots, a_c$) from knowledge of the r-independent variable categories ($b_1, \ldots, b_r$).

To illustrate the two error types, consider predicting category a_1 only from knowledge of its marginal distribution, $n_{.1}, \ldots, n_{.c}$. Clearly, $n_{.1}$ out of the N total cases is in category a_1, but exactly which $n_{.1}$ of the N cases is unknown. The probability of incorrectly identifying one of the N cases in category a_1 by chance alone is

$$\frac{N - n_{.1}}{N} \, .$$

Table 3.21 Notation for the cross-classification of two categorical variables, a_j for $j = 1, \ldots, c$ and b_i for $i = 1, \ldots, r$

	Variable a				
Variable b	a_1	a_2	$\cdots$	a_c	Total
b_1	n_{11}	n_{12}	$\cdots$	n_{1c}	$n_{1.}$
b_2	n_{21}	n_{22}	$\cdots$	n_{2c}	$n_{2.}$
$\vdots$	$\vdots$	$\vdots$	$\ddots$	$\vdots$	$\vdots$
b_r	n_{r1}	n_{r2}	$\cdots$	n_{rc}	$n_{r.}$
Total	$n_{.1}$	$n_{.2}$	$\cdots$	$n_{.c}$	N

Since there are $n_{.1}$ such classifications required, the number of expected incorrect classifications is

$$\frac{n_{.1}(N - n_{.1})}{N},$$

and, for all c categories of variable a, the number of expected errors of the first kind is

$$E_1 = \sum_{j=1}^{c} \frac{n_{.j}(N - n_{.j})}{N} .$$

Likewise, to predict $n_{11}, \ldots, n_{1c}$ from the independent category b_1, the probability of incorrectly classifying one of the $n_{1.}$ cases in cell n_{11} by chance alone is

$$\frac{n_{1.} - n_{11}}{n_{1.}} .$$

Since there are n_{11} such classifications required, the number of incorrect classifications is

$$\frac{n_{11}(n_{1.} - n_{11})}{n_{1.}},$$

and, for all cr cells, the number of expected errors of the second kind is

$$E_2 = \sum_{j=1}^{c} \sum_{i=1}^{r} \frac{n_{ij}(n_{i.} - n_{ij})}{n_{i.}} .$$

Goodman and Kruskal's t_a statistic is then defined as

$$t_a = \frac{E_1 - E_2}{E_1} .$$

An efficient computation form for Goodman and Kruskal's t_a is

$$t_a = \frac{N \displaystyle\sum_{i=1}^{r} \sum_{j=1}^{c} \frac{n_{ij}^2}{n_{i.}} - \sum_{j=1}^{c} n_{.j}^2}{N^2 - \displaystyle\sum_{j=1}^{c} n_{.j}^2} . \tag{3.3}$$

A computed value of t_a indicates the proportional reduction in prediction error given knowledge of the distribution of independent variable b over and above knowledge of only the distribution of dependent variable a. As defined, t_a is a point estimator of Goodman and Kruskal's population parameter τ_a for the population from which the sample of n cases was obtained. If variable b is considered, the dependent variable and variable a the independent variable, then Goodman and Kruskal's test statistic t_b and associated population parameter τ_b are analogously defined.

3.8.1 Example Analysis for t_a

For an example analysis employing Goodman and Kruskal's t_a measure of nominal association, consider the 3×4 contingency table in Table 3.22. For the data in Table 3.22, Goodman and Kruskal's t_a is

$$t_a = \frac{N \displaystyle\sum_{i=1}^{r} \sum_{j=1}^{c} \frac{n_{ij}^2}{n_{i.}} - \sum_{j=1}^{c} n_{.j}^2}{N^2 - \displaystyle\sum_{j=1}^{c} n_{.j}^2}$$

$$= \frac{90\left(\dfrac{5^2}{20} + \dfrac{0^2}{20} + \cdots + \dfrac{10^2}{40}\right) - (15^2 + 25^2 + 35^2 + 15^2)}{90^2 - (15^2 + 25^2 + 35^2 + 15^2)} = 0.1659\ .$$

Alternatively, errors of the first kind are

$$E_1 = \sum_{j=1}^{c} \frac{n_{.j}(N - n_{.j})}{N}$$

$$= \frac{15(90 - 15) + 25(90 - 25) + 35(90 - 35) + 15(90 - 15)}{90} = 64.4444\ ,$$

Table 3.22 Example 3×4 contingency table for Goodman and Kruskal's t_a and t_b measures

	Variable a				
Variable b	a_1	a_2	a_3	a_4	Total
b_1	5	0	15	0	20
b_2	5	5	15	5	30
b_3	5	20	5	10	40
Total	15	25	35	15	90

errors of the second kind are

$$E_2 = \sum_{j=1}^{c} \sum_{i=1}^{r} \frac{n_{ij}(n_{i.} - n_{ij})}{n_{i.}}$$

$$= \frac{5(20-5) + 0(20-0) + 15(20-15) + 0(20-0)}{20}$$

$$+ \frac{5(30-5) + 5(30-5) + 15(30-15) + 5(30-5)}{30}$$

$$+ \frac{5(40-5) + 20(40-20) + 5(40-5) + 10(40-10)}{40} = 53.75 \;,$$

and Goodman and Kruskal's t_a is

$$t_a = \frac{E_1 - E_2}{E_1} = \frac{64.4444 - 53.75}{64.4444} = 0.1659 \;.$$

The exact probability value of t_a under the null hypothesis is the sum of hypergeometric point probability values associated with t_a values equal to or greater than the observed value. For the data in Table 3.22, there are only $M = 3{,}453{,}501$ arrangements in the reference set of all permutations of cell frequencies, making an exact permutation analysis possible. The exact probability value of $t_a = 0.1659$ is $P = 0.3828$, i.e., the sum of the hypergeometric point probability values associated with values of $t_a = 0.1659$ or greater.

3.8.2 Example Analysis for t_b

Now consider variable b as the dependent variable. A convenient computing formula for t_b is

$$t_b = \frac{N \displaystyle\sum_{j=1}^{c} \sum_{i=1}^{r} \frac{n_{ij}^2}{n_{.j}} - \sum_{i=1}^{r} n_{i.}^2}{N^2 - \displaystyle\sum_{i=1}^{r} n_{i.}^2} \;.$$

Thus for the data in Table 3.22, Goodman and Kruskal's t_b is

$$t_b = \frac{90\left(\dfrac{5^2}{15} + \dfrac{0^2}{25} + \cdots + \dfrac{10^2}{40}\right) - (20^2 + 30^2 + 40^2)}{90^2 - (20^2 + 30^2 + 40^2)} = 0.2022 \;.$$

Alternatively, errors of the first kind are

$$E_1 = \sum_{j=1}^{c} \frac{n_{.j}(N - n_{.j})}{N} = \frac{20(90-20)+30(90-30)+40(90-40)}{90}$$

$$= 43.7778 \ ,$$

errors of the second kind are

$$E_2 = \sum_{j=1}^{c} \sum_{i=1}^{r} \frac{n_{ij}(n_{i.} - n_{ij})}{n_{i.}} = \frac{5(15-5)+5(15-5)+5(15-5)}{15}$$

$$+ \frac{0(25-0)+5(25-5)+20(25-20)}{25}$$

$$+ \frac{15(35-15)+15(35-15)+5(35-5)}{35}$$

$$+ \frac{0(15-0)+5(15-5)+10(15-10)}{15} = 46.0953 \ ,$$

and Goodman and Kruskal's t_b is

$$t_a = \frac{E_1 - E_2}{E_1} = \frac{57.7778 - 46.0953}{57.7778} = 0.2022 \ .$$

For the data in Table 3.22, there are only $M = 3{,}453{,}501$ arrangements in the reference set of all permutations of cell frequencies, making an exact permutation analysis feasible. The exact probability value of $t_b = 0.2022$ is $P = 0.5187$, i.e., the sum of the hypergeometric point probability values associated with values of $t_b = 0.2022$ or greater.

3.9 Monte Carlo Probability Values

Monte Carlo permutation procedures are essential when the number of arrangements of cell frequencies is large. For example, given the observed marginal frequency totals, the number of possible cell arrangements in Table 3.23 with $r = 3$ rows, $c = 5$ columns, and $N = 62$ observations is $M = 11{,}356{,}797$, rendering an exact permutation analysis impractical. A simple algorithmic procedure generates random arrangements of cell frequencies for any $r \times c$ contingency table, while preserving the observed row and column marginal frequency totals.[5]

[5] The procedure for generating random arrangements is easily generalized to multiway contingency tables [5].

Table 3.23 Example data for illustrating a Monte Carlo probability value

	Variable b					
Variable a	b_1	b_2	b_3	b_4	b_5	Total
a_1	7	3	5	5	1	21
a_2	0	7	6	6	3	22
a_3	1	2	6	6	4	19
Total	8	12	17	17	8	62

Table 3.24 Example of data illustrating the generation of random cell frequencies for an $r \times c$ contingency table

	Variable b							
Variable a	b_1	b_2	b_3	b_4	b_5	Total	Cumulative	Proportion
a_1	0	0	0	0	0	21	21	0.34
a_2	0	0	0	0	0	22	43	0.69
a_3	0	0	0	0	0	19	62	1.00
Total	8	12	17	17	8	62		
Cumulative	8	20	37	54	62			
Proportion	0.13	0.32	0.60	0.87	1.00			

Broadly speaking, the process employed to generate a random table begins with preserving the marginal frequency totals of the observed table and zeroes filling the cells in the table. The initial table also requires calculating the cumulative row and column marginal frequency totals and the row and column marginal proportions, the latter of which are used to construct a random table through multiple iterations [5]. Consider a detailed example using the data in Table 3.23.

Process for a Random Table

The first step in the procedure is to zero fill the cell frequencies, calculate the cumulative row and column marginal totals (Cumulative), and calculate the cumulative proportions of the marginal totals (Proportion), as in Table 3.24.

Next, two uniform random numbers $[0, 1)$ are generated, one for the row variable (UR) and one for the column variable (UC).[6] The first cumulative row proportion greater than random probability value UR determines the row of the cell to be selected, and the first cumulative column proportion greater than random probability value UC determines the column of the cell to be selected.

To illustrate the procedure, suppose that the initial random row and column probability values are $UR = 0.50$ and $UC = 0.40$, respectively. For the cumulative row proportions starting from the top of the column labeled Proportion, the first cumulative row proportion in Table 3.24 greater than $UR = 0.50$ is 0.69, so Row a_2 is selected. For the cumulative column proportions starting from the left of the row

[6] The notation $[0, 1)$ where 0 is preceded by a bracket and 1 is followed by a parenthesis means that the random numbers (UR) and (UC) will be equal to or greater than 0 and less than 1, i.e., $0 \leq UR < 1$ and $0 \leq UC < 1$.

Table 3.25 Example of data illustrating the generation of random cell frequencies for an $r \times c$ contingency table after selecting cell (2, 3)

	Variable b							
Variable a	b_1	b_2	b_3	b_4	b_5	Total	Cumulative	Proportion
a_1	0	0	0	0	0	21	21	0.34
a_2	0	0	1	0	0	21	42	0.69
a_3	0	0	0	0	0	19	61	1.00
Total	8	12	16	17	8	61		
Cumulative	8	20	36	53	61			
Proportion	0.13	0.33	0.59	0.87	1.00			

Table 3.26 Example of data illustrating the generation of random cell frequencies for an $r \times c$ contingency table after selecting cell (1, 4)

	Variable b							
Variable a	b_1	b_2	b_3	b_4	b_5	Total	Cumulative	Proportion
a_1	0	0	0	1	0	20	20	0.33
a_2	0	0	1	0	0	21	41	0.68
a_3	0	0	0	0	0	19	60	1.00
Total	8	12	16	16	8	60		
Cumulative	8	20	36	52	60			
Proportion	0.13	0.33	0.60	0.87	1.00			

labeled Proportion, the first cumulative column proportion greater than $UC = 0.40$ is 0.60, so Column b_3 is selected. With Row a_2 and Column b_3 selected, cell (2, 3) is increased by 1, the corresponding row and column marginal totals are decreased by 1 each, the new cumulative row and column marginal totals are determined, and the row and column cumulative proportions are recalculated, as in Table 3.25.

Now suppose that the next two random row and column probability values are $UR = 0.27$ and $UC = 0.72$, respectively. For the cumulative row proportions starting from the top of the column labeled Proportion, the first row proportion in Table 3.23 greater than $UR = 0.27$ is 0.34, so Row a_1 is selected. For the cumulative column proportions starting from the left of the row labeled Proportion, the first cumulative column proportion greater than $UC = 0.72$ is 0.87, so Column b_4 is selected. With Row a_1 and Column b_4 selected, cell (1, 4) is increased by 1, the corresponding row and column marginal totals are determined, and the row and column cumulative proportions are recalculated, as in Table 3.26.

Two new uniform random numbers (UR and UC) are obtained, and the process is repeated until all n observations have been randomly assigned. Depending on the random probability values UR and UC, the completed table will resemble Table 3.27. The process is repeated until the desired number of random tables (L) has been generated.

Table 3.27 Final random table with $r = 3$ rows, $c = 5$ columns, and $n = 62$ observations

	Variable b					
Variable a	b_1	b_2	b_3	b_4	b_5	Total
a_1	1	3	5	9	3	21
a_2	3	4	8	4	3	22
a_3	4	5	4	4	2	19
Total	8	12	17	17	8	62

Probability of a Test Statistic
If all M arrangements of the $n = 62$ observations in Table 3.25 occur with equal chance, the Monte Carlo probability value of an observed test statistic value computed on $L = 1{,}000{,}000$ randomly selected arrangements of the observed data with row and column marginals preserved for each arrangement is the proportion of random test statistic values equal to or greater than the observed test statistic value.

Example Analysis
To illustrate, for the frequency data in Table 3.21 with $n = 62$ observations, Pearson's chi-squared test statistic is $\chi^2 = 15.8060$, and the Monte Carlo probability value based on $L = 1{,}000{,}000$ random arrangements of cell frequencies is

$$P(\chi^2 \geq \chi_o^2 \mid \text{H}_0) = \frac{\text{number of } \chi^2 \text{ values } \geq \chi_o^2}{L} = \frac{44{,}300}{1{,}000{,}000} = 0.0443 ,$$

where χ_o^2 is the observed value of Pearson's χ^2. For comparison, the asymptotic probability value of $\chi^2 = 15.8060$ with $\nu = (r - 1)(c - 1) = (2)(4) = 8$ degrees of freedom is $P = 0.0452$. For this example analysis, the difference between the asymptotic and Monte Carlo probability values is only $P(0.0452 - 0.0443) = 0.0009$.

3.10 Summary

Chapter 3 presented permutation statistical methods, both exact and Monte Carlo, as applied to two cross-classified nominal-level (categorical) variables. The measures of relationship described in the chapter included Pearson's χ^2, Pearson's ϕ^2, Cramér's V^2, Fisher's exact probability test, and Goodman and Kruskal's t_a and t_b. A relatively new procedure for generating random contingency tables for Monte Carlo analyses was described and illustrated with an example analysis.

References

1. A. Agresti, B. Finlay, *Statistical Methods for the Social Sciences*, 4th edn. (Pearson, Edinburgh Gate, Essex, 2014)
2. G. Alberti, *From Data to Insights* (CRC Press, Boca Raton, FL, 2025)
3. R.A. Fisher, The logic of inductive inference (with discussion). J. R. Stat. Soc. **98**, 39–82 (1935)
4. L.A. Goodman, W.H. Kruskal, Measures of association for cross classifications. J. Am. Stat. Assoc. **49**, 732–764 (1954)
5. P.W. Mielke, K.J. Berry, J.E. Johnston, Resampling programs for multiway contingency tables with fixed marginal frequency totals. Psych. Rep. **101**, 18–24 (2007)
6. F. Mosteller, Association and estimation in contingency tables. J. Am. Stat. Assoc. **63**, 1–28 (1968)
7. S. Siegel, N.J. Castellan, *Nonparametric Statistics for the Behavioral Sciences*, 2nd edn. (McGraw–Hill, New York, 1988)

Chapter 4
Measures of Ordinal Association

Abstract This chapter presents exact and Monte Carlo permutation statistical methods applied to two cross-classified ordinal variables. Six sections describe and illustrate the application of permutation statistical methods to ordinal data. The first section introduces ordinal measures of association. Sections 4.2–4.6 present permutation applications for Kendall's τ_a measure, Kendall's τ_b measure, Stuart's τ_c measure, Goodman and Kruskal's γ measure, and Somers d_{yx} and d_{xy} measures, respectively.

Keywords Kendall's τ_a · Kendall's τ_b · Stuart's τ_c · Somers' d_{yx}

4.1 Introduction

This chapter presents exact and Monte Carlo permutation statistical methods applied to two cross-classified ordinal variables, commonly called ordinal contingency tables. Analyses of cross-classification tables for relationships between ordinal variables are of primary importance in basic research and are taught in every introductory course. Eight sections provide applications of permutation statistical methods to ordinal data. Section 4.1 provides an introduction to the analysis of cross-classified ordinal data. Section 4.2 provides a brief introduction to permutation statistical methods for ordinal data. Section 4.3 presents Kendall's S test statistic and Kendall's τ_a symmetric measure of ordinal association. Section 4.4 discusses Kendall's τ_b symmetric measure of ordinal association for $r \times c$ tables. Section 4.5 presents Stuart's τ_c symmetric measure of ordinal association. Section 4.6 presents Goodman and Kruskal's γ symmetric measure of ordinal association. Section 4.7 presents two asymmetric measures of ordinal association: Somers' d_{yx} and d_{xy}. Section 4.8 describes an iterative procedure for determining the maximum value of Kendall's S.

J. E. Johnston et al., *Permutation Statistical Methods: Measures of Relationship*,
SpringerBriefs in Statistics, https://doi.org/10.1007/978-3-032-26608-8_4

In general, for any permutation analysis of cross-classified data, there are four essential components:

1. The permutation test statistic, δ
2. The probability value, either exact or Monte Carlo, P
3. The exact expected value of test statistic δ, μ_δ
4. The chance-corrected measure of effect size, $\Re$

Just as there are many different conventional approaches for measuring the association between two ordinal variables, there are a variety of permutation methods for determining the association among ordinal variables. Two permutation methods are described and illustrated in this chapter. The first method is an exact probability analysis. The second is a Monte Carlo probability analysis.

A number of measures of association for two ordinal variables are based on pairwise comparisons of differences between rank scores. The test statistic S, as defined by Kendall in [4] and more extensively in [5], plays an important role in a variety of statistical measures where Kendall's test statistic is often expressed as $S = C - D$, where C and D indicate the number of concordant pairs and discordant pairs, respectively. Consider two ordinal variables cross-classified into an $r \times c$ contingency table, where r and c denote the number of rows and columns, respectively. Let $n_{i.}$, $n_{.j}$, and n_{ij} denote the row marginal frequency totals, column marginal frequency totals, and the number of objects in the ijth cell, respectively, for $i = 1, \ldots, r$ and $j = 1, \ldots, c$, and let n denote the total number of objects in the contingency table. Table 4.1 depicts a conventional notation for a typical $r \times c$ contingency table for two categorical variables, x_i for $i = 1, \ldots, r$ and y_j for $j = 1, \ldots, c$.

If x and y represent the row and column variables, respectively, there are $n(n-1)/2$ pairs of objects in the table that can be partitioned into five mutually exclusive, exhaustive types of pairs: concordant pairs, discordant pairs, pairs tied on variable x, pairs tied on variable y, and pairs tied on both variables x and y. For an $r \times c$ contingency table, concordant pairs (pairs of objects that are ranked in the same order on both variable x and variable y) are given by

$$C = \sum_{i=1}^{r-1} \sum_{j=1}^{c-1} n_{ij} \left(\sum_{k=i+1}^{r} \sum_{l=j+1}^{c} n_{kl} \right) ,$$

Table 4.1 Notation for the cross-classification of two categorical variables

	Variable y				
Variable x	1	2	$\cdots$	c	Total
1	n_{11}	n_{12}	$\cdots$	n_{1c}	$n_{1.}$
2	n_{21}	n_{22}	$\cdots$	n_{2c}	$n_{2.}$
$\vdots$	$\vdots$	$\vdots$	$\ddots$	$\vdots$	$\vdots$
r	n_{r1}	n_{r2}	$\cdots$	n_{rc}	$n_{r.}$
Total	$n_{.1}$	$n_{.2}$	$\cdots$	$n_{.c}$	n

discordant pairs (pairs of objects that are ranked in one order on variable x and the reverse order on variable y) are given by

$$D = \sum_{i=1}^{r-1} \sum_{j=1}^{c-1} n_{i,c-j+1} \left(\sum_{k=i+1}^{r} \sum_{l=1}^{c-j} n_{kl} \right) ,$$

pairs of objects tied on variable x are given by

$$T_x = \sum_{i=1}^{r} \sum_{j=1}^{c-1} n_{ij} \left(\sum_{k=j+1}^{c} n_{ik} \right) ,$$

pairs of objects tied on variable y are given by

$$T_y = \sum_{j=1}^{c} \sum_{i=1}^{r-1} n_{ij} \left(\sum_{k=i+1}^{r} n_{kj} \right) ,$$

and pairs of objects tied on variables x and y are given by

$$T_{xy} = \frac{1}{2} \sum_{i=1}^{r} \sum_{j=1}^{c} n_{ij} \left(n_{ij} - 1 \right) .$$

Then,

$$C + D + T_x + T_y + T_{xy} = \frac{n(n-1)}{2} .$$

Given C, D, T_x, T_y, and n, six measures of ordinal association are commonly defined, each having the same numerator, $S = C - D$, but different denominators [1]. The earliest of these pairwise measures was Kendall's τ_a [4]. Kendall's τ_a is a symmetrical measure of ordinal association that is most suitable when there are no tied pairs and is defined as the simple difference between the proportions of concordant and discordant pairs given by

$$\tau_a = \frac{C}{\frac{n(n-1)}{2}} - \frac{D}{\frac{n(n-1)}{2}} = \frac{C-D}{\frac{n(n-1)}{2}} = \frac{2S}{n(n-1)} .$$

Kendall's τ_b [5] extends τ_a to measure strong monotonicity[1] in contingency tables and is most appropriate when $r = c$. The denominator for τ_b is adjusted for the number of tied pairs for both variable x and variable y. Kendall's τ_b is given by

$$\tau_b = \frac{S}{\sqrt{(C + D + T_x)(C + D + T_y)}} .$$

Stuart's τ_c [12] modifies Kendall's τ_b for contingency tables where $r \neq c$ and is given by

$$\tau_c = \frac{2mS}{n^2(m - 1)} ,$$

where $m = \min(r, c)$.

Goodman and Kruskal's γ [3] is a symmetrical measure of weak monotonicity in which tied pairs of all types are ignored and given by

$$\gamma = \frac{S}{C + D} .$$

Somers' d_{yx} and d_{xy} [10] are asymmetric measures of ordinal association. Unlike the four symmetrical measures, τ_a, τ_b, τ_c, and γ, Somers' d_{yx} and d_{xy} measures depend on which variable, y or x, is considered the dependent variable. If variable y is the dependent variable, then

$$d_{yx} = \frac{S}{C + D + T_y} ,$$

and if variable x is the dependent variable, then

$$d_{xy} = \frac{S}{C + D + T_x} .$$

4.2 Permutation Statistical Methods

For an exact permutation analysis of an $r \times c$ contingency table, it is necessary to calculate the selected measure of ordinal association for the observed cell frequencies and exhaustively enumerate all possible arrangements of the n objects

---

[1] Monotonicity refers to the relationship between variables x and y. For strong monotonicity, as variable x increases (decreases), variable y also increases (decreases). For weak monotonicity, as variable x increases (decreases), variable y either increases (decreases) or stays the same.

in the rc cells, given the observed marginal frequency distributions. For each arrangement in the reference set of all permutations of cell frequencies, a measure of ordinal association, say, T, and the exact hypergeometric point probability value under the null hypothesis, $p(n_{ij} \mid n_{i.}, n_{.j}, n)$, are calculated. A hypergeometric point probability value is

$$p(n_{ij} \mid n_{i.}, n_{.j}, n) = \frac{\left(\prod_{i=1}^{r} n_{i.}! \right) \left(\prod_{j=1}^{c} n_{.j}! \right)}{n! \prod_{i=1}^{r} \prod_{j=1}^{c} n_{ij}!} ,$$

where n_{ij} is an observed cell frequency for $i = 1, \ldots r$ and $j = 1, \ldots, c$, $n_{i.}$ is the ith of r row marginal frequency totals summed over all columns, $n_{.j}$ is the jth of c column marginal frequency totals summed over all rows, and n is the total of all n_{ij} values for $i = 1, \ldots, r$ and $j = 1, \ldots, c$ [2, 9, p. 258]. If T_{o} denotes the value of the observed test statistic, the exact one-sided upper- and lower-tail probability (P) values of T_{o} are the sums of the $p(n_{ij} \mid n_{i.}, n_{.j}, n)$ values associated with the T values computed on all possible arrangements of cell frequencies that are equal to or greater than T_{o} when T_{o} is positive and equal to or less than T_{o} when T_{o} is negative, respectively. Thus the hypergeometric probability of T_{o} when T is positive is

$$P = \sum_{k=1}^{M} \Psi(T_k)\, p(n_{ij} \mid n_{i.}, n_{.j}, n) ,$$

where

$$\Psi(T_k) = \begin{cases} 1 & \text{if } T_k \geq T_{\text{o}} , \\ 0 & \text{otherwise} , \end{cases}$$

and the exact hypergeometric probability value of T_{o} when T is negative is

$$P = \sum_{k=1}^{M} \Psi(T_k)\, p(n_{ij} \mid n_{i.}, n_{.j}, n) ,$$

where

$$\Psi(T_k) = \begin{cases} 1 & \text{if } T_k \leq T_{\text{o}} , \\ 0 & \text{otherwise} . \end{cases}$$

When the number of possible arrangements of cell frequencies is very large, exact tests are impractical and Monte Carlo permutation statistical methods become necessary. Monte Carlo permutation statistical methods generate a random sample of all possible arrangements of cell frequencies, given the marginal frequency distributions. The Monte Carlo one-sided upper- and lower-tail probability values of statistic T are simply the proportions of the T values computed on the randomly selected arrangements of cell frequencies that are equal to or greater than T_o when T_o is positive and equal to or less than T_o when T_o is negative, respectively. Thus, the Monte Carlo probability of T_o when T is positive is

$$P(T \geq T_o \,|\, H_0) = \frac{\text{number of } T \text{ values } \geq T_o}{L} \, ,$$

and when T is negative is given by

$$P(T \leq T_o \,|\, H_0) = \frac{\text{number of } T \text{ values } \leq T_o}{L} \, ,$$

where L is the number of random arrangements of the data.

4.3 Kendall's τ_a Measure

Kendall's τ_a measure of ordinal association [4],

$$\tau_a = \frac{2S}{n(n-1)} \, ,$$

was originally designed to measure the association between two sets of untied rank scores, where the two sets of rank scores are labeled x and y, although the rank scores are usually represented in an $r \times c$ contingency table where $n_{i.} = n_{.j} = 1$ for $i = 1, \ldots, r$ and $j = 1, \ldots, c$.

4.3.1 Example Analysis

For an example analysis employing Kendall's τ_a measure of ordinal association, consider the data in Table 4.2, where $n = 20$ bivariate observations have been cross-classified into a 3×3 ordered contingency table. The number of observations is kept small for illustration purposes. For the data in Table 4.2, the number of concordant pairs is

$$C = \sum_{i=1}^{r-1} \sum_{j=1}^{c-1} n_{ij} \left(\sum_{k=i+1}^{r} \sum_{l=j+1}^{c} n_{kl} \right) = (6)(2+1+1+5) + (2)(1+5)$$

$$+ (2)(1+5) + (2)(5) = 88 \, ,$$

the number of discordant pairs is

$$D = \sum_{i=1}^{r-1} \sum_{j=1}^{c-1} n_{i,c-j+1} \left(\sum_{k=i+1}^{r} \sum_{l=1}^{c-j} n_{kl} \right) = (0)(2+2+1+1)$$
$$+ (2)(2+1) + (1)(1+1) + (2)(1) = 10 ,$$

the number of pairs tied on variable x is

$$T_x = \sum_{i=1}^{r} \sum_{j=1}^{c-1} n_{ij} \left(\sum_{k=j+1}^{c} n_{ik} \right) = (6)(2+0) + (2)(0) + (2)(2+1)$$
$$+ (2)(1) + (1)(1+5) + (1)(5) = 31 ,$$

the number of pairs tied on variable y is

$$T_y = \sum_{j=1}^{c} \sum_{i=1}^{r-1} n_{ij} \left(\sum_{k=i+1}^{r} n_{kj} \right) = (6)(2+1) + (2)(1) + (2)(2+1)$$
$$+ (2)(1) + (0)(1+5) + (1)(5) = 33 ,$$

the number of pairs tied on both variables x and y is

$$T_{xy} = \frac{1}{2} \sum_{i=1}^{r} \sum_{j=1}^{c} n_{ij} \left(n_{ij} - 1 \right) = \frac{1}{2}\big[(6)(6-1) + (2)(2-1)$$
$$+ (0)(0-1) + (2)(2-1) + (2)(2-1) + (1)(1-1) + (1)(1-1)$$
$$+ (1)(1-1) + (5)(5-1)\big] = 28 ,$$

Kendall's S is $C - D = 88 - 10 = +78$, and Kendall's τ_a is[2]

$$\tau_a = \frac{2S}{n(n-1)} = \frac{2(+78)}{20(20-1)} = +0.4105 .$$

4.3.2 Permutation Probability

For the data in Table 4.2, there are only $M = 412$ arrangements in the reference set of all permutations of cell frequencies, making an exact permutation analysis

[2] Note that tied values are calculated, but Kendall's τ_a ignores all ties.

Table 4.2 Rank-score data for $n = 20$ bivariate observations cross-classified on ordinal variables x and y

	Variable y			
Variable x	1	2	3	Total
1	6	2	0	8
2	2	2	1	5
3	1	1	5	7
Total	9	5	6	20

feasible. If all arrangements of the data in Table 4.2 occur with equal chance, the exact probability of Kendall's S under the null hypothesis is the sum of the hypergeometric point probability values associated with $S = +78$ or greater. Based on the hypergeometric probability distribution, the exact probability value is $P = 0.0026$.

4.3.3 Asymptotic Probability

In testing the significance of the association between paired ranks, it is more efficient to apply a test directly to S than to τ_a as the number of pairs, $n(n-1)/2$, is a constant [2]. Kendall's S statistic is asymptotically distributed $N(0, 1)$[3] under the null hypothesis with a mean of zero and variance given by

$$\sigma_S^2 = \frac{n(n-1)(2n+5)}{18}$$

as $n \to \infty$. Since the normal distribution is an approximation to the discrete sampling distribution of S, a correction for continuity should be applied. For the data in Table 4.2, the standard normal deviate with continuity correction applied is

$$z = \frac{|S| - 1}{\sqrt{n(n-1)(2n+5)/18}} = \frac{78 - 1}{\sqrt{20(20-1)[(2)(20)+5]/18}} = +3.8069 ,$$

yielding an approximate two-tail probability value of $P = 1.4072 \times 10^{-4}$, under the assumptions of independence, random sampling, and normality. For this example analysis, the difference between the exact and asymptotic probability values is only $P(0.0026 - 1.4072 \times 10^{-4}) = 2.4595 \times 10^{-3}$.

[3] $N(0, 1)$ is an abbreviation for a normal distribution.

4.4 Kendall's τ_b Measure

Because tied values occur in many datasets, Kendall's τ_a measure of ordinal association is less than satisfactory for many problems, as it ignores any tied values. For this reason, Kendall developed τ_b [5], an alternative to τ_a given by

$$\tau_b = \frac{S}{\sqrt{(C+D+T_x)(C+D+T_y)}} ,$$

which incorporates tied values on variables x and y (T_x and T_y, respectively).

4.4.1 Example Analysis

For an example analysis employing Kendall's τ_b measure of ordinal association, consider the data in Table 4.3 in which $n = 72$ bivariate observations have been cross-classified into a 3×5 ordered contingency table. For the data in Table 4.3, the number of concordant pairs is

$$C = \sum_{i=1}^{r-1} \sum_{j=1}^{c-1} n_{ij} \left(\sum_{k=i+1}^{r} \sum_{l=j+1}^{c} n_{kl} \right) = (8)(2+8+5+5+5+3+7+7)$$
$$+ (4)(8+5+5+3+7+7) + \cdots + (8)(7+7) + (5)(7) = 855 ,$$

the number of discordant pairs is

$$D = \sum_{i=1}^{r-1} \sum_{j=1}^{c-1} n_{i,c-j+1} \left(\sum_{k=i+1}^{r} \sum_{l=1}^{c-j} n_{kl} \right)$$
$$= (3)(3+2+8+5+4+5+3+7) + (5)(3=2+8+4+5+3)$$
$$+ \cdots + (8)(4+5) + (2)(4) = 541 ,$$

Table 4.3 Rank-score data for $n = 72$ bivariate observations cross-classified on ordinal variables x and y

	Variable y					
Variable x	1	2	3	4	5	Total
1	8	4	3	5	3	23
2	3	2	8	5	5	23
3	4	5	3	7	7	26
Total	15	11	14	17	15	72

the number of pairs tied on variable x is

$$T_x = \sum_{i=1}^{r} \sum_{j=1}^{c-1} n_{ij} \left(\sum_{k=j+1}^{c} n_{ik} \right) = (8)(4+3+5+3) + (4)(3+5+3) + \cdots + (3)(7+7) + (7)(7) = 668 \ ,$$

the number of pairs tied on variable y is

$$T_y = \sum_{j=1}^{c} \sum_{i=1}^{r-1} n_{ij} \left(\sum_{k=i+1}^{r} n_{kj} \right)$$
$$= (8)(3+4) + (3)(4) + (4)(2+5) + (2)(5) + \cdots + (3)(5+7) + (5)(7) = 329 \ ,$$

Kendall's S is $C - D = 866 - 541 = +314$, and Kendall's τ_b is

$$\tau_b = \frac{S}{\sqrt{(C+D+T_x)(C+D+T_y)}}$$
$$= \frac{+314}{\sqrt{(855+541+668)(855+541+329)}} = +0.1664 \ .$$

4.4.2 Permutation Probability

For the data in Table 4.3, there are $M = 70{,}148{,}145$ arrangements in the reference set of all permutations of cell frequencies. Therefore, an exact test is not practical and a Monte Carlo procedure based on $L = 1{,}000{,}000$ random arrangements of cell frequencies is used. The Monte Carlo probability of τ_b under the null hypothesis is the number of τ_b values equal to or greater than $\tau_b = +0.1664$; in this case, the Monte Carlo probability value is

$$P(\tau_b \geq \tau_o) = \frac{\text{number of } \tau_b \text{ values } \geq \tau_o}{L} = \frac{48{,}600}{1{,}000{,}000} = 0.0486 \ ,$$

where τ_o is the observed value of τ_b.

While an exact permutation analysis is not practical for the data in Table 4.3, it is not impossible. The exact probability value of Kendall's τ_b under the null hypothesis is the sum of the hypergeometric point probability values associated with values of $\tau_b = +0.1664$ or greater. Based on the hypergeometric probability distribution, the exact upper-tail probability value conditioned on the reference set of all $M = 70{,}148{,}145$ arrangements of cell frequencies is $P = 0.0488$. For this example analysis, the difference between the exact and Monte Carlo probability values is only $P(0.0488 - 0.0486) = 0.0002$.

4.4.3 Asymptotic Probability

In testing the significance of the association between paired ranks, it is more efficient to apply a test directly to S than to τ_b as the number of pairs, $n(n-1)/2$, is a constant [2]. Kendall's S statistic is asymptotically distributed $N(0, 1)$ under the null hypothesis with a mean of zero and variance

$$\sigma_S^2 = \frac{n(n-1)(2n+5)}{18}$$

as $n \to \infty$. Since the normal distribution is an approximation to the discrete sampling distribution of S, a correction for continuity should be applied. For the data in Table 4.3, the standard normal deviate with continuity correction applied is

$$z = \frac{|S| - 1}{\sqrt{n(n-1)(2n+5)/18}} = \frac{314 - 1}{\sqrt{72(72-1)[(2)(72)+5]/18}} = +1.5216 \, ,$$

yielding an approximate two-tail probability value of $P = 0.1282$, under the assumptions of independence, random sampling, and normality. For this example analysis, the difference between the Monte Carlo and asymptotic probability values is $P(0.1282 - 0.0488) = 0.0794$.

4.5 Stuart's τ_c Measure

Kendall's τ_b is a strongly monotonic measure of ordinal association, i.e., for every ordered category increase in variable x, there is expected to be an ordered category increase in variable y. Consequently, τ_b can only achieve limits of ± 1 for contingency tables where $r = c$ and the marginal frequency distributions are identical. More specifically, τ_b cannot generally attain values of ± 1 because of the Cauchy inequality:

> The square of the sum of the products of two sets will be equal to or less than the product of the squared sums of two sets.

More formally, for variables x and y,

$$\left(\sum_{i=1}^{n} x_i y_y \right)^2 \le \sum_{i=1}^{n} x_i^2 \sum_{i=1}^{n} y_i^2 \, .$$

Consequently, the numerator of τ_b will be equal to or less than the denominator, permitting τ_b to attain ± 1 only when all the observations are concentrated on one of the two principal diagonals of the contingency table. If no marginal frequency is to be zero, τ_b can attain ± 1 only for a square ($r = c$) contingency table with identical marginal frequency distributions.

Thus Kendall's τ_b is not the most appropriate measure of ordinal association for the 3×5 contingency table in Table 4.3. To correct for this limitation, Stuart proposed τ_c for contingency tables where $r \neq c$, given by

$$\tau_c = \frac{2mS}{n^2(m-1)} ,$$

where $m = \min(r, c)$ [12], which in a classic case of Stigler's [11] law of eponymy, is often erroneously referred to as Kendall's τ_c. Stuart showed that if n is a multiple of m and $r = c$ with identical marginal frequency distributions such that all observations fall on the diagonal of the contingency table and all cell frequencies are equal, the maximum value of S is given by

$$S_{\max} = \frac{n^2(m-1)}{2m} . \tag{4.1}$$

Then if $n = m$,

$$\frac{n^2(m-1)}{2m} = \frac{n^2(n-1)}{2n} = \frac{n(n-1)}{2} .$$

However, if n is not a multiple of m, the expression in Eq. (4.1) remains an upper bound that cannot be attained. It follows that Stuart's τ_c can sometimes attain, and for large n, can almost always attain ± 1.

4.5.1 Example Analysis

For an example analysis employing Stuart's τ_c measure of ordinal association, consider the data in Table 4.3, replicated for convenience in Table 4.4, in which $n = 72$ bivariate observations have been cross-classified into a 3×5 ordered contingency table. For the data in Table 4.4, the number of concordant pairs is

$$C = \sum_{i=1}^{r-1} \sum_{j=1}^{c-1} n_{ij} \left(\sum_{k=i+1}^{r} \sum_{l=j+1}^{c} n_{kl} \right) = (8)(2+8+5+5+5+3+7+7)$$
$$+ (4)(8+5+5+3+7+7) + \cdots + (8)(7+7) + (5)(7) = 855 ,$$

the number of discordant pairs is

$$D = \sum_{i=1}^{r-1} \sum_{j=1}^{c-1} n_{i,c-j+1} \left(\sum_{k=i+1}^{r} \sum_{l=1}^{c-j} n_{kl} \right)$$
$$= (3)(3+2+8+5+4+5+3+7) + (5)(3+2+8+4+5+3)$$
$$+ \cdots + (8)(4+5) + (2)(4) = 541 \ ,$$

Kendall's S is $C - D = 855 - 541 = +314$, and Stuart's τ_c is

$$\tau_c = \frac{2mS}{n^2(m-1)} = \frac{2(3)(+314)}{72^2(3-1)} = +0.1817 \ .$$

For this example analysis, there is little difference between Kendall's $\tau_b = +0.1664$ and Stuart's $\tau_c = +0.1817$, i.e., $0.1817 - 0.1664 = 0.0153$.

4.5.2 Permutation Probability

For the data in Table 4.4, there are $M = 70{,}148{,}145$ arrangements in the reference set of all permutations of cell frequencies, making an exact permutation analysis possible. The exact probability value of Stuart's τ_c under the null hypothesis is the sum of the hypergeometric point probability values associated with values of $\tau_c = +0.1817$ or greater. Based on the hypergeometric probability distribution, the exact probability value is $P = 0.0600$.

4.5.3 Asymptotic Probability

In testing the significance of the association between paired ranks, it is more efficient to apply a test directly to S than to τ_c as the number of pairs, $n(n-1)/2$, is a constant [2]. Kendall's S statistic is asymptotically distributed $N(0, 1)$ under the null hypothesis with a mean of zero and variance given by

$$\sigma_S^2 = \frac{n(n-1)(2n+5)}{18}$$

Table 4.4 Rank-score data for Stuart's τ_c with $n = 72$ bivariate observations

	Variable y					
x	1	2	3	4	5	Total
1	8	4	3	5	3	23
2	3	2	8	5	5	23
3	4	5	3	7	7	26
Total	15	11	14	17	15	72

as $n \to \infty$. Since the normal distribution is an approximation to the discrete sampling distribution of S, a correction for continuity should be applied. For the data in Table 4.4, the standard normal deviate with continuity correction applied is

$$z = \frac{|S| - 1}{\sqrt{n(n-1)(2n+5)/18}} = \frac{314 - 1}{\sqrt{72(72-1)[(2)(72)+5]/18}} = +1.5216 \,,$$

yielding an approximate two-tail probability value of $P = 0.1282$, under the assumptions of independence, random sampling, and normality. For this example analysis, the difference between the exact and asymptotic probability values is $P(0.1282 - 0.0600) = 0.0682$.

4.6 Goodman and Kruskal's γ Measure

In 1954, Goodman and Kruskal developed a new measure of association for two ordinal variables called gamma (γ) [3]. Gamma is a proportional-reduction-in-error measure of ordinal association that is based solely on the untied pairs, C and D, and given by

$$\gamma = \frac{S}{C + D} \,.$$

4.6.1 Example Analysis

For an example analysis employing Goodman and Kruskal's γ measure of ordinal association, consider the data in Table 4.5, in which $n = 75$ bivariate observations have been cross-classified into a 3×5 ordered contingency table. For the data in

Table 4.5 Rank-score data for Goodman and Kruskal's γ with $n = 75$ bivariate observations

	Variable y					
Variable x	1	2	3	4	5	Total
1	9	5	3	1	1	19
2	4	7	8	3	2	24
3	2	4	8	8	10	32
Total	15	16	19	12	13	75

Table 4.5, the number of concordant pairs is

$$C = \sum_{i=1}^{r-1} \sum_{j=1}^{c-1} n_{ij} \left(\sum_{k=i+1}^{r} \sum_{l=j+1}^{c} n_{kl} \right)$$

$$= (9)(7+8+3+2+4+8+8+10) + (5)(8+3+2+8+8+10)$$
$$+ \cdots + (8)(8+10) + (3)(10) = 1{,}202 \; ,$$

the number of discordant pairs is

$$D = \sum_{i=1}^{r-1} \sum_{j=1}^{c-1} n_{i,c-j+1} \left(\sum_{k=i+1}^{r} \sum_{l=1}^{c-j} n_{kl} \right)$$

$$= (1)(4+7+8+3+2+4+8+8) + (1)(4+7+8+2+4+8)$$
$$+ \cdots + (8)(2+4) + (7)(2) = 306 \; ,$$

Kendall's S is $C - D = 1{,}202 - 306 = +896$, and Goodman and Kruskal's γ is

$$\gamma = \frac{S}{C+D} = \frac{+896}{1{,}202 + 306} = +0.5942 \; .$$

4.6.2 Permutation Probability

For the data in Table 4.5, there are only $M = 68{,}161{,}105$ arrangements in the reference set of all permutations of cell frequencies, making an exact permutation analysis feasible. The exact probability value of Goodman and Kruskal's γ under the null hypothesis is the sum of the hypergeometric point probability values associated with values of $\gamma = +0.5942$ or greater. Based on the hypergeometric probability distribution, the probability value is only $P = 1.5180 \times 10^{-6}$.

4.6.3 Asymptotic Probability

In testing the significance of the association between paired ranks, it is more efficient to apply a test directly to S than to γ as the number of pairs, $n(n-1)/2$, is a constant [2]. Kendall's S statistic is asymptotically distributed $N(0, 1)$ under the null hypothesis with a mean of zero and variance given by

$$\sigma_S^2 = \frac{n(n-1)(2n+5)}{18}$$

as $n \rightarrow \infty$. Since the normal distribution is an approximation to the discrete sampling distribution of S, a correction for continuity should be applied. For the data in Table 4.5, the standard normal deviate with continuity correction applied is

$$z = \frac{|S| - 1}{\sqrt{n(n-1)(2n+5)/18}} = \frac{896 - 1}{\sqrt{75(75-1)[(2)(75)+5]/18}} = +4.0940 ,$$

yielding an approximate two-tail probability value of $P = 4.2400 \times 10^{-5}$, under the assumptions of independence, random sampling, and normality. For this example analysis, the difference between the exact and asymptotic probability values is only $P(1.5180 \times 10^{-4} - 4.2400 \times 10^{-5}-) = 1.0940 \times 10^{-4}$.

4.7 Somers' d_{yx} and d_{xy} Measures

In 1962, Somers proposed two asymmetric measures of ordinal association given by

$$d_{yx} = \frac{S}{C + D + T_y} ,$$

where T_y is the number of pairs tied on variable y, and

$$d_{xy} = \frac{S}{C + D + T_x} ,$$

where T_x is the number of pairs tied on variable x [10].

4.7.1 Example Analysis

For an example analysis employing Somers' d_{xy} measure of ordinal association, with x the dependent variable, consider the data in Table 4.6 in which $n = 42$ bivariate observations have been cross-classified into a 3×4 ordered contingency

Table 4.6 Rank-score data for Somers' d_{xy} measure with $n = 42$ bivariate observations

	Variable y				
Variable x	1	2	3	4	Total
1	6	5	6	1	18
2	4	3	5	2	14
3	2	1	4	3	10
Total	12	9	15	6	42

table. For the data in Table 4.6, the number of concordant pairs is

$$C = \sum_{i=1}^{r-1} \sum_{j=1}^{c-1} n_{ij} \left(\sum_{k=i+1}^{r} \sum_{l=j+1}^{c} n_{kl} \right) = (6)(3+5+2+1+4+3)$$
$$+ (5)(5+2+4+3) + (6)(2+3) + (4)(1+4+3)$$
$$+ (3)(4+3) + (5)(3) = 276 ,$$

the number of discordant pairs is

$$D = \sum_{i=1}^{r-1} \sum_{j=1}^{c-1} n_{i,c-j+1} \left(\sum_{k=i+1}^{r} \sum_{l=1}^{c-j} n_{kl} \right) = (1)(4+3+5+2+1+4)$$
$$+ (6)(4+3+2+1) + (5)(4+2) + (2)(2+1+4)$$
$$+ (5)(2+1) + (3)(2) = 144 ,$$

the number of pairs tied on variable x is

$$T_x = \sum_{i=1}^{r} \sum_{j=1}^{c-1} n_{ij} \left(\sum_{k=j+1}^{c} n_{ik} \right) = (6)(5+6+1) + (5)(6+1)$$
$$+ (6)(1) + (4)(3+5+2) + (3)(5+2) + (5)(2) + (2)(1+4+3)$$
$$+ (1)(4+3) + (4)(3) = 219 ,$$

Kendall's S is $C - D = 276 - 144 = +132$, and Somers' d_{xy} is

$$d_{xy} = \frac{S}{C + D + T_x} = \frac{+132}{276 + 144 + 219} = +0.2066 .$$

4.7.2 Permutation Probability

For the data in Table 4.6, there are only $M = 72{,}143$ arrangements in the reference set of all permutations of cell frequencies, making an exact permutation analysis feasible. However, in this case, consider over-sampling via a Monte Carlo procedure. Over-sampling occurs quite often in the permutation literature as commercial permutation programs often do not have an exact probability option. In addition, Monte Carlo over-sampling can be much more efficient than an exact procedure as the hypergeometric point probability values need not be calculated for each arrangement of the data.

The Monte Carlo probability value of Somers' d_{xy} is simply the proportion of randomly selected d_{xy} values equal to or greater than $d_{xy} = +0.2066$. If d_{o} denotes the observed value of d_{xy}, the Monte Carlo probability based on $L = 1{,}000{,}000$ random arrangements of the cell frequencies is

$$P(d_{xy} \geq d_{\text{o}} \,|\, \text{H}_0) = \frac{\text{number of } d_{xy} \text{ values } \geq d_{\text{o}}}{L} = \frac{55{,}581}{1{,}000{,}000} = 0.0556 \,.$$

4.7.3 Asymptotic Probability

In testing the significance of the association between paired ranks, it is more efficient to apply a test directly to S than to d_{xy} as the number of pairs, $n(n-1)/2$, is a constant under permutation [2]. Kendall's S statistic is asymptotically distributed $N(0, 1)$ under the null hypothesis with a mean of zero and variance given by

$$\sigma_S^2 = \frac{n(n-1)(2n+5)}{18}$$

as $n \to \infty$. Since the normal distribution is an approximation to the discrete sampling distribution of S, a correction for continuity should be applied. For the data in Table 4.6, the standard normal deviate with continuity correction applied is

$$z = \frac{|S| - 1}{\sqrt{n(n-1)(2n+5)/18}} = \frac{132 - 1}{\sqrt{42(42-1)[(2)(42)+5]/18}} = +1.4197 \,,$$

yielding an approximate two-tail probability value of $P = 0.1556$, under the assumptions of independence, random sampling, and normality. For this example analysis, the difference between the Monte Carlo and asymptotic probability values is $P(0.1556 - 0.0556) = 0.1000$.

4.7.4 Example Analysis

For an example analysis employing Somers' d_{yx} measure of ordinal association, with y the dependent variable, consider the data in Table 4.6 in which $n = 42$ bivariate observations have been cross-classified into a 3×4 contingency table. For the ordered data in Table 4.6, the number of concordant pairs is $C = 276$, the number of discordant pairs is $D = 144$, the number of pairs tied on variable y is

$$T_y = \sum_{j=1}^{c} \sum_{i-1}^{r-1} n_{ij} \left(\sum_{k=i+1}^{r} \right) = (6)(4+2) + (4)(2) + (5)(3+1)$$
$$+ (3)(1) + (6)(5+4) + (5)(4) + (1)(2+3) + (2)(3) = 152 \,,$$

Kendall's S is $C - D = 176 - 144 = +132$, and Somers' d_{yx} is

$$d_{yx} = \frac{S}{C + D + T_y} = \frac{+132}{276 + 144 + 152} = +0.2308 \; .$$

4.7.5 Permutation Probability

For the data in Table 4.6, there are only $M = 72{,}143$ arrangements in the reference set of all permutations of cell frequencies, making an exact permutation analysis feasible. The exact probability value of $d_{yx} = +0.2308$ is $P = 0.0452$, that is, the sum of the hypergeometric point probability values associated with values of $d_{yx} = +0.2308$ or greater.

4.7.6 Asymptotic Probability

In testing the significance of the association between paired ranks, it is more efficient to apply a test directly to S than to d_{yx} as the number of pairs, $n(n-1)/2$, is a constant under permutation [2]. Kendall's S statistic is asymptotically distributed $N(0, 1)$ under the null hypothesis with a mean of zero and variance given by

$$\sigma_S^2 = \frac{n(n-1)(2n+5)}{18}$$

as $n \to \infty$. Since the normal distribution is an approximation to the discrete sampling distribution of S, a correction for continuity should be applied. For the data in Table 4.6, the standard normal deviate with continuity correction applied is

$$z = \frac{|S| - 1}{\sqrt{n(n-1)(2n+5)/18}} = \frac{132 - 1}{\sqrt{42(42-1)[(2)(42)+5]/18}} = +1.4197 \; ,$$

yielding an approximate two-tail probability value of $P = 0.1556$, under the assumptions of independence, random sampling, and normality. For this example analysis, the difference between the exact and asymptotic probability values is $P(0.1556 - 0.0452) = 0.1104$.

4.8 Maximum Value for S

The S statistic, introduced by Kendall in [4], is the principal component in a number of measures of ordinal association, including Kendall's τ_a [4], Kendall and

Babington Smith's W [6], Kendall and Babington Smith's u [7], Whitfield's S [13], Kendall's τ_b [5], Stuart's τ_c [12], Somers' d_{yx} and d_{xy} [10], Kim's $d_{y.x}$ and $d_{x.y}$ [8], and Wilson's e [14].

In order for any of these measures to attain values of -1 or $+1$, certain conditions are requisite. For example, given an $r \times c$ contingency table, for many of these measures, a value of $+1$ can be attained only if $r = c$. If the observed table is not square, say, an $r \neq c$ contingency table, none of these measures can attain a value of $+1$. Thus, intermediate values, say, $\tau_b = +0.60$, may be uninterpretable, since the maximum value of S is unknown.

4.8.1 Structure of Ordinal Measures

A measure of bivariate ordinal association typically consists of a numerator and a denominator. The numerator is composed of a statistic, such as Kendall's S, that provides a raw assessment of the amount of association. The principal function of the denominator is to scale the numerator statistic, while assuming some model of association, thus providing a measure of ordinal association between the limits of -1 and $+1$.

Consider two sets of cell frequencies with identical marginal distributions. One set consists of the observed cell frequencies and reflects the degree of ordinal association in the table, given the marginal distributions. The other set consists of generated cell frequencies that reflect the maximum possible degree of ordinal association in the table, given the marginal distributions. The first set is the basis for the numerator of the measure, and the second set is the basis for the denominator of the measure.

The problem with measures based on Kendall's S, such as τ_a, τ_b, τ_c, γ, and others, is that the numerator is based on the observed cell frequencies and marginals, but the denominator is based on some theoretical, idealized model of association, often positing that variable x is a necessary and sufficient condition of variable y. It is simply not possible to design a closed-form denominator that will be applicable to cross-classification tables of all sizes and any marginal distributions. However, there is another approach.

The solution is to construct a denominator from the observed data that scales the numerator statistic between the limits of -1 and $+1$. In brief, if Kendall's S is the numerator statistic, then the denominator needs to be the maximum value of S, given the marginal distributions of the observed contingency table.

In this section, a simple algorithmic procedure is described to establish the maximum value of Kendall's S for any size contingency table with any row or column frequency distributions. The algorithm is based on the marginal distributions of the observed table and generates cell frequencies that, when analyzed, provide the maximum value of S for the denominator of the measure.

4.8.2 Notation

To introduce the algorithmic procedure, consider two ordinal variables cross-classified into an $r \times c$ contingency table, where r and c denote the number of rows and columns, respectively. Let $n_{i.}$, $n_{.j}$, and n_{ij} denote the row marginal frequency totals, the column marginal frequency totals, and number of objects in the ijth cell, respectively, for $i = 1, \ldots, r$ and $j = 1, \ldots, c$, and let n denote the total number of objects in the table. If x and y represent the row and column variables, respectively, there are $n(n-1)/2$ pairs of objects in the table that can be partitioned into five mutually exclusive, exhaustive types of pairs: concordant pairs, discordant pairs, pairs tied on variable x, pairs tied on variable y, and pairs tied on both variables x and y.

For an $r \times c$ contingency table, the number of concordant pairs (pairs of objects that are ranked in the same order on variable x and variable y) is given by

$$C = \sum_{i=1}^{r-1} \sum_{j=1}^{c-1} n_{ij} \left(\sum_{k=i+1}^{r} \sum_{l=j+1}^{c} n_{kl} \right) ,$$

the number of discordant pairs (pairs of objects that are ranked in one order on variable x and the reverse order on variable y) is given by

$$D = \sum_{i=1}^{r-1} \sum_{j=1}^{c-1} n_{i,c-j+1} \left(\sum_{k=i+1}^{r} \sum_{l=1}^{c-j} n_{kl} \right) ,$$

the number of pairs of objects tied on variable x but differing on variable y is given by

$$T_x = \sum_{i=1}^{r} \sum_{j=1}^{c-1} n_{ij} \left(\sum_{k=j+1}^{c} n_{ik} \right) ,$$

the number of pairs of objects tied on variable y but differing on variable x is given by

$$T_y = \sum_{j=1}^{c} \sum_{i=1}^{r-1} n_{ij} \left(\sum_{k=i+1}^{r} n_{kj} \right) ,$$

and the number of pairs of objects tied on both variable x and variable y is given by

$$T_{xy} = \frac{1}{2} \sum_{i=1}^{r} \sum_{j=1}^{c} n_{ij}(n_{ij} - 1) .$$

Then,

$$C + D + T_x + T_y + T_{xy} = \frac{n(n-1)}{2} .$$

4.8.3 *Procedure*

In this section, the algorithmic procedure for finding the maximum value of S is described and illustrated step by step. For a positive value of S, the procedure is as follows. Proceeding from Cell $(1, 1)$ to Cell (r, c) by columns, enter into each cell the smaller of the two corresponding cell marginals and subtract that value from each corresponding marginal and the table total. For this illustration, a 3×3 contingency table is used with row marginals $\{10, 20, 30\}$ and column marginals $\{20, 15, 25\}$. The procedure begins with an empty initial table containing the marginal frequencies (Table 4.7).

Step 1: Using the marginals from the initial table for Cell $(1, 1)$, enter $\min(10, 20) = 10$ (row marginal) into Cell $(1, 1)$ and subtract 10 from the table total and the corresponding marginals (n, $n_{1.}$, and $n_{.1}$), producing Table 4.8.

Step 2: Using the adjusted marginals from Table 4.8 for Cell $(2, 1)$, enter $\min(20, 10) = 10$ (column marginal) into Cell $(2, 1)$ and subtract 10 from the adjusted table total and the corresponding adjusted marginals (n, $n_{2.}$, and $n_{.1}$), producing Table 4.9.

Step 3: Using the adjusted marginals from Table 4.9 for Cell $(3, 1)$, enter $\min(30, 0) = 0$ (column marginal) into Cell $(3, 1)$ and subtract 0 from the adjusted

Table 4.7 Initial 3×3 table with empty cells and row and column marginals given

x/y	1	2	3	Sum
1				10
2				20
3				30
Sum	20	15	25	60

Table 4.8 Introduction of the first cell frequency in Cell $(1, 1)$

x/y	1	2	3	Sum
1	**10**			**0**
2				20
3				30
Sum	**10**	15	25	**50**

Table 4.9 Introduction of the second cell frequency in Cell $(2, 1)$

x/y	1	2	3	Sum
1	10			0
2	**10**			**10**
3				30
Sum	**0**	15	25	**40**

Table 4.10 Introduction of the third cell frequency in Cell $(3, 1)$

x/y	1	2	3	Sum
1	10			0
2	10			10
3	**0**			**30**
Sum	**0**	15	25	**40**

Table 4.11 Introduction of the fourth cell frequency in Cell $(1, 2)$

x/y	1	2	3	Sum
1	10	**0**		**0**
2	10			10
3	0			30
Sum	0	**15**	25	**40**

Table 4.12 Introduction of the fifth cell frequency in Cell $(2, 2)$

x/y	1	2	3	Sum
1	10	0		0
2	10	**10**		**0**
3	0			30
Sum	0	**5**	25	**30**

Table 4.13 Introduction of the sixth cell frequency in Cell $(3, 2)$

x/y	1	2	3	Sum
1	10	0		0
2	10	10		0
3	0	**5**		**25**
Sum	0	**0**	25	**25**

table total and the corresponding adjusted marginals (n, $n_{3.}$, and $n_{.1}$), producing Table 4.10.

Step 4: Using the adjusted marginals from Table 4.10 for Cell $(1, 2)$, enter $\min(0, 15) = 0$ (row marginal) into Cell $(1, 2)$ and subtract 0 from the adjusted table total and the corresponding adjusted marginals (n, $n_{1.}$, and $n_{.2}$), producing Table 4.11.

Step 5: Using the adjusted marginals from Table 4.11 for Cell $(2, 2)$, enter $\min(10, 15) = 10$ (row marginal) into Cell $(2, 2)$ and subtract 10 from the adjusted table total and the corresponding adjusted marginals (n, $n_{2.}$, and $n_{.2}$), producing Table 4.12.

Step 6: Using the adjusted marginals from Table 4.12 for Cell $(3, 2)$, enter $\min(30, 5) = 5$ (column marginal) into Cell $(3, 2)$ and subtract 5 from the adjusted table total and the corresponding adjusted marginals (n, $n_{3.}$, and $n_{.2}$), producing Table 4.13.

Step 7: Using the adjusted marginals from Table 4.13 for Cell $(1, 3)$, enter $\min(0, 25) = 0$ (row marginal) into Cell $(1, 3)$ and subtract 0 from the adjusted table total and the corresponding adjusted marginals (n, $n_{1.}$, and $n_{.3}$), producing Table 4.14.

Table 4.14 Introduction of the seventh cell frequency in Cell (1, 3)

x/y	1	2	3	Sum
1	10	0	**0**	**0**
2	10	10		0
3	0	5		25
Sum	0	0	**25**	**25**

Table 4.15 Introduction of the eighth cell frequency in Cell (2, 3)

x/y	1	2	3	Sum
1	10	0	0	0
2	10	10	**0**	**0**
3	0	5		25
Sum	0	0	**25**	**25**

Table 4.16 Introduction of the ninth cell frequency in Cell (3, 3)

x/y	1	2	3	Sum
1	10	0	0	0
2	10	10	0	0
3	0	5	**25**	**0**
Sum	0	0	**0**	**0**

Table 4.17 The maximized 3×3 contingency table

x/y	1	2	3	Sum
1	10	0	0	10
2	10	10	0	20
3	0	5	25	30
Sum	20	15	25	60

Step 8: Using the adjusted marginals from Table 4.14 for Cell (2, 3), enter $\min(0, 25) = 0$ (row marginal) into Cell (2, 3) and subtract 0 from the adjusted table total and the corresponding adjusted marginals (n, $n_{2.}$, and $n_{.3}$), producing Table 4.15.
Step 9: Using the adjusted marginals from Table 4.15 for Cell (3, 3), enter $\min(25, 25) = 25$ (row or column marginal) into Cell (3, 3) and subtract 25 from the adjusted table total and the corresponding adjusted marginals (n, $n_{3.}$, and $n_{.3}$), producing Table 4.16.
The final table with the marginals reconstituted is in Table 4.17.

Note that the described procedure is for any positive value of S. For negative values of S, the procedure is the same except for the order of the cells. For negative values of S, begin with the right uppermost corner, Cell$(1, c)$, instead of the left uppermost corner, Cell (1, 1), and proceed down each column from right to left, across the table.

4.8.4 Example Analysis

In this section, the procedure is demonstrated with a worked example. Kendall's τ_b is selected to illustrate the procedure. Consider the 3×3 contingency table in Table 4.18 with the same marginal distributions as Table 4.17. The cell frequencies in Table 4.18 are completely arbitrary and have been selected to illustrate the procedure.

For the observed data in Table 4.18, the number of concordant pairs is

$$C = \sum_{i=1}^{r-1} \sum_{j=1}^{c-1} n_{ij} \left(\sum_{k=i+1}^{r} \sum_{l=j+1}^{c} n_{kl} \right) = (5)(6+4+6+19) + (3)(4+19)$$

$$+ (10)(6+19) + (6)(19) = 608 \; ,$$

the number of discordant pairs is

$$D = \sum_{i=1}^{r-1} \sum_{j=1}^{c-1} n_{i,c-j+1} \left(\sum_{k=i+1}^{r} \sum_{l=1}^{c-j} n_{kl} \right) = (2)(10+6+5+6)$$

$$+ (3)(10+5) + (4)(5+6) + (6)(5) = 173 \; ,$$

the number of pairs tied on variable x is

$$T_x = \sum_{i=1}^{r} \sum_{j=1}^{c-1} n_{ij} \left(\sum_{k=j+1}^{c} n_{ik} \right) = (5)(3+2) + (3)(2) + (10)(6+4)$$

$$+ (6)(4) + (5)(6+19) + (6)(19) = 394 \; ,$$

the number of pairs tied on variable y is

$$T_y = \sum_{j=1}^{c} \sum_{i=1}^{r-1} n_{ij} \left(\sum_{k=i+1}^{r} n_{kj} \right) = (5)(10+5) + (10)(5) + (3)(6+6)$$

$$+ (6)(6) + (2)(4+19) + (4)(19) = 319 \; ,$$

Table 4.18 Contingency data to illustrate the maximizing procedure

x/y	1	2	3	Sum
1	5	3	2	10
2	10	6	4	20
3	5	6	19	30
Sum	20	15	25	60

Kendall's S is $C - D = 608 - 173 = +435$, and Kendall's τ_b is

$$\tau_b = \frac{S}{\sqrt{(C + D + T_x)(C + D + T_y)}}$$

$$= \frac{+435}{\sqrt{(608 + 173 + 394)(608 + 173 + 319)}} = +0.3826 \; .$$

4.8.5 Comparison with Maximized Cell Frequencies

The results can be compared with an analysis of the maximized cell frequencies in Table 4.17, where the number of concordant pairs is

$$C = \sum_{i=1}^{r-1} \sum_{j=1}^{c-1} n_{ij} \left(\sum_{k=i+1}^{r} \sum_{l=j+1}^{c} n_{kl} \right) = (10)(10 + 0 + 5 + 25)$$

$$+ (10)(5 + 25) + (10)(25) = 950 \; ,$$

the number of discordant pairs is $D = 0$, the number of pairs tied on variable x is

$$T_x = \sum_{i=1}^{r} \sum_{j=1}^{c-1} n_{ij} \left(\sum_{k=j+1}^{c} n_{ik} \right) = (10)(0 + 0) + (0)(0) + (10)(10 + 0)$$

$$+ (0)(5 + 25) + (5)(25) = 225 \; ,$$

the number of pairs tied on variable y is

$$T_y = \sum_{j=1}^{c} \sum_{i=1}^{r-1} n_{ij} \left(\sum_{k=i+1}^{r} n_{kj} \right) = (10)(10 + 0) + (10)(0) + (0)(10 + 5)$$

$$+ (10)(5) + (10)(0) + (0)(0 + 25) + (0)(25) = 150 \; ,$$

Kendall's maximized S is $S_{\max} = C - D = 950 - 0 = +950$, and the maximized coefficient of bivariate ordinal association (T) is

$$T = \frac{S}{S_{\max}} = \frac{+435}{950} = +0.4579 \; .$$

4.8.6 Conclusion

Conventional measures of ordinal association based on Kendall's S measure underestimate the degree of association due to denominators based on a theoretical model instead of the data structure of the observed table. Using an algorithmic procedure, Kendall's S is computed to the maximum allowed by the observed data, providing coefficients of ordinal association that, in general, are considerably larger than conventional coefficients. For this example analysis, the maximized T coefficient is considerably larger than Kendall's t_b coefficient, with a difference of $0.4579 - 0.3826 = 0.0753$.

4.9 Summary

Chapter 4 presented permutation statistical methods, both exact and Monte Carlo, as applied to two cross-classified ordinal-level (ranked) variables. The measures of relationship described in the chapter included Kendall's S, Kendall's τ_a and τ_b, Stuart's τ_c, Goodman and Kruskal's γ, and Somers' d_{yx} and d_{xy}. A relatively new procedure for determining the maximum value of Kendall's S statistic was described and illustrated with an example analysis.

References

1. K.J. Berry, J.E. Johnston, P.W. Mielke, Exact and resampling probability values for measures associated with ordered R by C contingency tables. Psychol. Rep. **99**, 231–238 (2006)
2. E.J. Burr, The distribution of Kendall's score S for a pair of tied rankings. Biometrika **47**, 151–171 (1960)
3. L.A. Goodman, W.H. Kruskal, Measures of association for cross classifications. J. Am. Stat. Assoc. **49**, 732–764 (1954)
4. M.G. Kendall, A new measure of rank correlation. Biometrika **30**, 81–93 (1938)
5. M.G. Kendall, *Rank Correlation Methods* (Griffin, London, 1948)
6. M.G. Kendall, B. Babington Smith, The problem of m rankings. Annals Math. Stat. **10**, 275–287 (1939)
7. M.G. Kendall, B. Babington Smith, On the method of paired comparisons. Biometrika **31**, 324–345 (1940)
8. J.O. Kim, Predictive measures of ordinal association. Am. J. Soc. **76**, 891–907 (1971)
9. P.W. Mielke, K.J. Berry, *Permutation Methods: A Distance Function Approach* (Springer, New York, 2001)
10. R.H. Somers, A new asymmetric measure of association for ordinal variables. Am. Sociol. Rev. **27**, 799–811 (1962)
11. S.M. Stigler, Stigler's law of eponymy, in *Science and Social Structure: A Festschrift for Robert K. Merton*, ed. by T.F. Gieryn (New York Academy of Sciences, New York, 1980), pp. 147–157

12. A. Stuart, The estimation and comparison of strengths of association in contingency tables. Biometrika **40**, 105–110 (1953)
13. J.W. Whitfield, Rank correlation between two variables, one of which is ranked, the other dichotomous. Biometrika **34**, 292–296 (1947)
14. T.P. Wilson, Measures of association for bivariate ordinal hypotheses, in *Measurement in the Social Sciences: Theories and Strategies*, ed. by H.M. Blalock (Aldine, Chicago, 1974), pp. 327–342

Chapter 5
Measures of Correlation

Abstract This chapter presents exact and Monte Carlo permutation statistical methods applied to two correlated variables. Permutation methods as applied to Pearson's product–moment correlation coefficient, the intraclass correlation coefficient, Spearman's rank-order correlation coefficient, Spearman's footrule measure of agreement, and Kendall and Babington Smith's coefficient of concordance are described and illustrated.

Keywords Correlation · Concordance · Analysis of ranks · Intraclass correlation

5.1 Introduction

This chapter presents exact and Monte Carlo permutation statistical methods applied to two correlated variables. Five sections describe and illustrate various measures of correlation. Section 5.2 introduces and illustrates permutation statistical methods applied to Pearson's conventional product–moment correlation coefficient. Section 5.3 presents data analyses employing the intraclass correlation coefficient. Section 5.4 presents measures of association based on Spearman's rank-order correlation coefficient. Section 5.5 presents data analyses employing Spearman's footrule measure of inter-rater agreement. Section 5.6 describes permutation methods applied to Kendall and Babington Smith's coefficient of concordance.

In general, for any permutation analysis of cross-classified data, there are four essential components:

1. The permutation test statistic, δ
2. The probability value, either exact or Monte Carlo, P
3. The exact expected value of test statistic δ, μ_δ
4. The chance-corrected measure of effect size, $\Re$

J. E. Johnston et al., *Permutation Statistical Methods: Measures of Relationship*, SpringerBriefs in Statistics, https://doi.org/10.1007/978-3-032-26608-8_5

5.2 Pearson's Product–Moment Correlation

Ordinary least squares (OLS) regression with a single predictor is a popular statistical measure of the degree of correlation between two interval variables, usually denoted as x and y. The assumption of normality comes into play when the null hypothesis is tested by conventional means. Permutation statistical methods do not assume normality and, therefore, are often more useful than conventional statistical methods, especially when the sample size is small or the population distribution is heavy tailed [10]. Let r_{xy} denote the Pearson product–moment correlation coefficient for variables x and y given by

$$r_{xy} = \frac{n \sum_{i=1}^{n} x_i y_i - \sum_{i=1}^{n} x_i \sum_{i=1}^{n} y_i}{\sqrt{\left[n \sum_{i=1}^{n} x_i^2 - \left(\sum_{i=1}^{n} x_i \right)^2 \right] \left[n \sum_{i=1}^{n} y_i^2 - \left(\sum_{i=1}^{n} y_i \right)^2 \right]}} ,$$

where n is the number of bivariate measurements. The conventional test of significance for r_{xy} is

$$t = r_{xy} \sqrt{\frac{n-2}{1-r_{xy}^2}} ,$$

which is distributed as Student's t with $n - 2$ degrees of freedom, under the assumptions of independence, homogeneity, linearity, and normality.

More popular than simple OLS regression and correlation is multiple OLS regression with p predictors, $x_1, x_2, \ldots, x_p$. Let $R_{y.x_1, x_2, \ldots, x_p}$ indicate the multiple correlation coefficient for variables y and $x_1, x_2, \ldots, x_p$ given by

$$R^2_{x_1, x_2, \ldots, x_p} = \boldsymbol{\beta}' \mathbf{r}_y ,$$

where $\boldsymbol{\beta}'$ is the transposed vector of standardized regression weights and $\mathbf{r}_y$ is the vector of zero-order correlation coefficients of y with $x_1, x_2, \ldots, x_p$. The conventional test of significance for $R^2_{x_1, x_2, \ldots, x_p}$ is

$$F = \frac{(n-p-1) R^2_{y.x_1, x_2, \ldots, x_p}}{p(1 - R^2_{y.x_1, x_2, \ldots, x_p})} ,$$

which is distributed as Snedecor's F with p and $n - p - 1$ degrees of freedom, under the assumptions of independence, homogeneity, linearity, and normality.

5.2.1 Univariate Example

For an example analysis of Pearson's product-moment correlation coefficient, consider the bivariate data in Table 5.1 for $n = 11$ objects. For the data in Table 5.1 with $n = 11$,

$$\sum_{i=1}^{n} x_i = 186,\ \sum_{i=1}^{n} x_i^2 = 3{,}448,\ \sum_{i=1}^{n} y_i = 87,\ \sum_{i=1}^{n} y_i^2 = 889,\ \sum_{i=1}^{n} x_i y_i = 1,681,$$

and Pearson's product–moment correlation coefficient is

$$r_{xy} = \frac{n\sum_{i=1}^{n} x_i y_i - \sum_{i=1}^{n} x_i \sum_{i=1}^{n} y_i}{\sqrt{\left[n\sum_{i=1}^{n} x_i^2 - \left(\sum_{i=1}^{n} x_i\right)^2\right]\left[n\sum_{i=1}^{n} y_i^2 - \left(\sum_{i=1}^{n} y_i\right)^2\right]}}$$

$$= \frac{(11)(1{,}681) - (186)(87)}{\sqrt{[(11)(3{,}448) - (186)^2][(11)(889) - (87)^2]}} = +0.8509\ .$$

5.2.2 Permutation Probability

An exact permutation analysis requires random shuffles of either the x or the y values. Although with this small example there are only

$$M = n! = 11! = 39{,}916{,}800$$

Table 5.1 Bivariate correlation data on $n = 11$ objects

Object	x	y
1	11	2
2	22	8
3	14	3
4	11	4
5	10	5
6	26	17
7	20	10
8	17	12
9	14	6
10	17	9
11	24	11

arrangements in the reference set of all permutations of the data, making an exact permutation analysis feasible, Monte Carlo permutation methods are generally preferred for correlation analyses since $n!$ is usually a very large number, e.g., $13! = 6{,}227{,}020{,}800$. Let r_o indicate the observed value of r_{xy}. Then based on $L = 1{,}000{,}000$ random arrangements of the data under the null hypothesis, there are 861 r_{xy} values equal to or greater than $r_o = +0.8509$, yielding a Monte Carlo probability of

$$P(r_{xy} \mid \mathrm{H}_0) = \frac{\text{number of } r_{xy} \text{ values } \geq r_o}{L} = \frac{861}{1{,}000{,}000} = 8.6100 \times 10^{-4} \; .$$

5.2.3 *Asymptotic Probability*

For comparison, for the data in Table 5.1,

$$t = r_{xy}\sqrt{\frac{n-2}{1-r_{xy}^2}} = +0.8509\sqrt{\frac{11-2}{1-(0.8509)^2}} = +4.8591 \; ,$$

and the two-tail probability value of $r_{xy} = +0.8509$ based on Student's t distribution with $n-2 = 11-2 = 9$ degrees of freedom is $P = 8.9690 \times 10^{-4}$. For this example analysis, the difference between the Monte Carlo and asymptotic probability values is only $P(8.9690 \times 10^{-4} - 8.6100 \times 10^{-4}) = 3.5900 \times 10^{-5}$.

5.2.4 *Multivariate Example*

For a multivariate example of OLS correlation, consider the small dataset with $p = 2$ predictors in Table 5.2 where variable y is weight in pounds, variable x_1 is height in inches, and variable x_2 is age in years for $n = 12$ elementary school children. For the multivariate data in Table 5.2, the multiple correlation coefficient is $R^2_{y.x_1,x_2} = 0.7301$ (henceforth, simply R^2).

5.2.5 *Permutation Probability*

An exact permutation analysis of multiple correlation requires random shuffles of either the x values or the y values. It is important to note that the predictor variables must be shuffled as a unit, i.e., $x_1, \ldots, x_p$. Otherwise, a researcher may end up with a combination of predictor variables that make no sense, i.e., 4-year-old child, married, with two children. Thus, it is advisable to simply shuffle the y values. Even

Table 5.2 Multivariate correlation data on $n = 12$ children

Child	x_1	x_2	y
1	48	7	55
2	41	13	53
3	47	10	59
4	60	10	78
5	56	9	66
6	61	12	69
7	49	8	60
8	58	11	73
9	50	9	57
10	51	7	58
11	56	10	70
12	54	11	79

with this very small example, there are

$$M = n! = 12! = 479{,}001{,}600$$

possible arrangements of the data, making an exact permutation analysis impractical. Based on $L = 1{,}000{,}000$ random arrangements of the data, the Monte Carlo probability of $R^2 = 0.7301$ is

$$P\big(R^2 \geq R_{\text{o}}^2 \,|\, \text{H}_0\big) = \frac{\text{number of } R^2 \text{ values } \geq R_{\text{o}}^2}{L} = \frac{2{,}370}{1{,}000{,}000}$$
$$= 2.3700 \times 10^{-3} ,$$

where R_{o}^2 is the observed value of R^2.

5.2.6 Asymptotic Probability

For comparison, for the multivariate data in Table 5.2,

$$F = \frac{(n-p-1)R^2_{y.x_1,\,x_2,\,\ldots,\,x_p}}{p(1-R^2_{y.x_1,\,x_2,\,\ldots,\,x_p})} = \frac{(11-2-1)(0.7301)}{2(1-0.7301)} = 12.1728,$$

and the probability value of $R^2 = 0.7301$ based on Snedecor's F distribution with $p = 2$ and $n - p - 1 = 12 - 2 - 1 = 9$ degrees of freedom is $P = 2.7569 \times 10^{-3}$. For this example analysis, the difference between the Monte Carlo and asymptotic probability values is only $P(2.7569 \times 10^{-3} - 2.3700 \times 10^{-3}) = 3.8690 \times 10^{-4}$.

5.3 Intraclass Correlation

There is an extensive and controversial literature on the intraclass correlation coefficient and its uses [5, 7]. The standard reference is Haggard [2], although it has been heavily criticized for both its exposition and its statistical accuracy [6].

The intraclass correlation coefficient is most often used for measuring the level of agreement among judges. The coefficient represents concordance, where $+1$ indicates perfect agreement and 0 indicates no agreement. While the maximum value of the intraclass correlation coefficient is $+1$, the minimum is $-1/(k-1)$, where k is the number of judges. Thus, for $k = 2$ judges, the lower limit is -1, but for $k = 3$ judges the lower limit is $-1/2$, for $k = 4$ judges the lower limit is $-1/3$, for $k = 5$ judges the lower limit is $-1/4$, and so on, approaching zero as the number of judges increases.

5.3.1 *Example Analysis*

For an example analysis employing the intraclass correlation coefficient, consider the data in Table 5.3 with $k = 3$ judges and $n = 5$ objects. The number of observations is kept small for illustration purposes.

The intraclass correlation coefficient is [6]

$$r_{\text{I}} = \frac{k \sum_{i=1}^{n} \left(\bar{x}_i - \bar{\bar{x}}\right)^2}{n(k-1)S_x^2} - \frac{1}{k-1} ,$$

where n is the number of objects, k is the number of judges, and the sample variance is given by[1]

Table 5.3 Data for an intraclass correlation with $n = 5$ subjects and $k = 3$ judges

	Judge		
Subject	x_1	x_2	x_3
1	12	10	8
2	15	11	7
3	9	9	6
4	6	5	4
5	8	5	5

[1] Note that this is the sample variance (S_x^2), not the estimated population variance (s_x^2). So, in this case, the sum of squares is divided by nk, not $nk - 1$ as is typical and symbolized by an uppercase letter S.

$$S_x^2 = \frac{1}{nk} \sum_{i=1}^{nk} \left(x_i - \bar{\bar{x}}\right)^2 \ ,$$

where $\bar{\bar{x}}$ is the grand mean over all nk observations.

For the data in Table 5.3, the grand mean is

$$\bar{\bar{x}} = \frac{1}{nk} \sum_{i=1}^{nk} x_i = \frac{12 + 10 + 8 + \cdots + 8 + 5 + 5}{(5)(3)} = \frac{120}{15} = 8.00 \ ,$$

the sample variance is

$$\begin{aligned} S_x^2 &= \frac{1}{nk} \sum_{i=1}^{nk} \left(x_i - \bar{\bar{x}}\right)^2 \\ &= \frac{(12 - 8.00)^2 + (10 - 8.00)^2 + \cdots + (5 - 8.00)^2 + (5 - 8.00)^2}{(5)(3)} \\ &= \frac{132}{15} = 8.80 \ , \end{aligned}$$

the sum of squares is calculated in Table 5.4, and the intraclass correlation coefficient is

$$r_{\text{I}} = \frac{k \sum_{i=1}^{n} \left(\bar{x}_i - \bar{\bar{x}}\right)^2}{n(k-1)S_x^2} - \frac{1}{k-1} = \frac{(3)(26.00)}{5(3-1)(8.80)} - \frac{1}{3-1} = +0.3864 \ .$$

Table 5.5 contains the analysis of variance source table for the data in Table 5.3. The intraclass correlation coefficient is distributed as Snedecor's F with $k - 1$ and $(k-1)(n-1)$ degrees of freedom. Given the source table in Table 5.5, the probability of $F = 11.43$ with $k - 1 = 3 - 1 = 2$ and $(k - 1)(n - 1) = (3 - 1)(5 - 1) = 8$ degrees of freedom is $P = 0.0045$.

Table 5.4 Means, grand mean, and sum of squares for the data in Table 5.3

Object	x_1	x_2	x_3	$\bar{x}$	$\bar{\bar{x}}$	$(\bar{x} - \bar{\bar{x}})^2$
1	12	10	8	10	8.00	4.00
2	15	11	7	11	8.00	9.00
3	9	9	6	8	8.00	0.00
4	6	5	4	5	8.00	9.00
5	8	5	5	6	8.00	4.00
Sum						26.00

Table 5.5 Analysis of variance source table for the data in Table 5.3 with $k = 3$ judges and $n = 5$ subjects

Source	*SS*	*df*	*MS*	*F*
Between S	78.00	4	19.50	
Within S	54.00	10	5.40	
Factor A	40.00	2	20.00	11.43
Error ($A \times S$)	14.00	8	1.75	
Total	132.00	14		

5.3.2 *A Permutation Analysis*

Permutation analyses are completely data dependent and do not depend on random sampling and/or fixed- or random-effects models. For the data in Table 5.3, there are only

$$M = \left(a!\right)^n = \left(3!\right)^5 = 7{,}776$$

arrangements in the reference set of all permutations of the data, making an exact permutation analysis possible. If r_o is the observed value of r_I, the exact probability value of $r_\text{I} = +0.3864$ is

$$P\left(r_\text{I} \geq r_\text{o} \,|\, \text{H}_0\right) = \frac{\text{number of } r_\text{I} \text{ values } \geq r_\text{o}}{M} = \frac{24}{7{,}776} = 0.0031 \ .$$

In this example analysis, the difference between the Monte Carlo and asymptotic probability values is only $P(0.0045 - 0.0031) = 0.0014$.

5.4 Spearman's Rank-Order Correlation

Consider two rankings of n objects consisting of the first n integers, and let x_i and y_i for $i = 1, \ldots, n$ denote the first and second rankings, respectively. A popular measure of correlation between the two rankings is Spearman's rank-order correlation coefficient, ρ, given by

$$\rho = 1 - \frac{6 \displaystyle\sum_{i=1}^{N} d_i^2}{n(n^2 - 1)} \ ,$$

where $d_i = x_i - y_i$ for $i = 1, \ldots, n$ [8].

5.4.1 Example Analysis

For an example analysis using Spearman's rank-order correlation coefficient, consider the data in Table 5.6 with $n = 13$ objects and two sets of scores, x and y. For the data in Table 5.6, the columns headed x and y contain the raw scores, the columns headed r_x and r_y contain the corresponding rank scores, the column headed d contains the signed differences between r_x and r_y, and the column headed d^2 contains the squared rank differences. For the data in Table 5.6, Spearman's rank-order correlation coefficient is

$$\rho = 1 - \frac{6\sum_{i=1}^{n} d_i^2}{n(n^2-1)} = 1 - \frac{6(166)}{13(13^2-1)} = +0.5440 \ .$$

5.4.2 Permutation Probability

Because there are $M = n! = 13! = 6{,}227{,}020{,}800$ possible arrangements of the data in Table 5.6, an exact permutation test is not practical and a Monte Carlo probability procedure based on $L = 1{,}000{,}000$ random arrangements of the rank scores is used. There are 28,975 values equal to or greater than $\rho = +0.5440$. If all arrangements in the reference set of the L random arrangements in Table 5.6 occur

Table 5.6 Example data for Spearman's rank-order correlation with $n = 13$ objects

Pair	x	y	r_x	r_y	d	d^2
1	65	70	5	1	+4	16
2	68	64	4	4	0	0
3	55	42	8	11	−3	9
4	44	60	10	6	+4	16
5	72	50	3	9	−6	36
6	36	58	12	7	+5	25
7	60	68	6	2	+4	16
8	73	63	2	5	−3	9
9	32	35	13	13	0	0
10	40	55	11	8	+3	9
11	75	66	1	3	−2	4
12	56	38	7	12	−5	25
13	50	45	9	10	−1	1
Sum						166

with equal chance, the approximate Monte Carlo probability of $\rho = +0.5440$ under the null hypothesis is

$$P(\rho \geq \rho_o \,|\, H_0) = \frac{\text{number of } \rho \text{ values } \geq \rho_o}{L} = \frac{28{,}975}{1{,}000{,}000} = 0.0290 \,,$$

where ρ_o is the observed value of Spearman's ρ.

5.4.3 Asymptotic Probability

Spearman's ρ is distributed as Student's t with $n - 2$ degrees of freedom. For the data in Table 5.6,

$$t = \rho\sqrt{\frac{n-2}{1-\rho^2}} = +0.5440\sqrt{\frac{13-2}{1-(0.5440)^2}} = 2.1500 \,,$$

yielding a probability value of $P = 0.0546$. For this example analysis, the difference between the Monte Carlo and asymptotic probability values is $P(0.0546 - 0.0290) = 0.0256$.

5.5 Spearman's Footrule Measure

A 1906 article by Spearman contained a new measure of rank-order correlation that Spearman named "the footrule" [9]. Consider two rankings of n objects consisting of the first n integers, and let x_i and y_i for $i = 1, \ldots, n$ denote the first and second rankings, respectively. Then Spearman's footrule is given by

$$\mathcal{R} = 1 - \frac{3\sum_{i=1}^{n} |x_i - y_i|}{n^2 - 1} \,.$$

Spearman's $\mathcal{R}$ is a chance-corrected measure of agreement between variables x and y, attaining a maximum value of $+1$ when x_i is identical to y_i for $i = 1, \ldots, n$ and no tied values are present [1]. However, if $y_i = n - x_i + 1$ for $i = 1, \ldots, n$, then $\mathcal{R} = -0.5$ when n is odd and

$$\mathcal{R} = -0.5\left(1 + \frac{3}{n^2 - 1}\right)$$

when n is even [3]. Consequently, $\mathcal{R}$ cannot attain a minimum value of -1, except when $n = 2$.

5.5.1 Example Analysis

For an example analysis employing Spearman's footrule measure, consider the paired-rank data in Table 5.7 with $n = 12$ objects. For the data in Table 5.7, Spearman's footrule is

$$\mathcal{R} = 1 - \frac{3\sum_{i=1}^{n} |x_i - y_i|}{n^2 - 1} = 1 - \frac{3(26)}{12^2 - 1} = +0.4545 \,,$$

indicating approximately 45% agreement above that expected by chance.

5.5.2 Permutation Probability

Because there are $M = n! = 12! = 479{,}001{,}600$ arrangements in the reference set of all permutations of the x and y rank scores in Table 5.6, an exact permutation test is not practical and a Monte Carlo probability procedure based on $L = 1{,}000{,}000$ random arrangements of cell frequencies is used. There are 19,115 values equal to or greater than $\mathcal{R} = +0.4545$. If all arrangements of the rank scores in Table 5.6

Table 5.7 Data for Spearman's footrule rank correlation with $n = 12$ objects and two rankings, x and y

Pair	x	y	$x - y$	$\lvert x - y\rvert$
1	7	5	+2	2
2	3	1	+2	2
3	4	2	+2	2
4	11	10	+1	1
5	9	12	−3	3
6	8	9	−1	1
7	5	8	−3	3
8	6	7	−1	1
9	12	6	+6	6
10	10	11	−1	1
11	1	3	−2	2
12	2	4	−2	2
Sum				26

occur with equal chance, the Monte Carlo probability of $\mathcal{R} = +0.4545$ under the null hypothesis is

$$P(\mathcal{R} \geq \mathcal{R}_\text{o} \,|\, \text{H}_0) = \frac{\text{number of } \mathcal{R} \text{ values } \geq \mathcal{R}_\text{o}}{L} = \frac{19{,}115}{1{,}000{,}000} = 0.0191 \ ,$$

where $\mathcal{R}_\text{o}$ is the observed value of $\mathcal{R}$.

5.5.3 *Asymptotic Probability*

It is easily shown that Kendall's S statistic can be defined in terms of Spearman's footrule when there are no tied values. For the data in Table 5.6, Kendall's S, defined in terms of Spearman's $\mathcal{R}$, is

$$S = \frac{(n-1)[n + 2\mathcal{R}(n+1) - 2]}{6}$$
$$= \frac{(12-1)[12 + (2)(0.4545)(12+1) - 2]}{6} = 40.00 \ .$$

It is well known that Kendall's S is distributed $N(0, 1)$ under the null hypothesis with a mean of zero and variance

$$\sigma_S^2 = \frac{n(n-1)(2n+5)}{18}$$

as $n \to \infty$. Since the normal distribution is an approximation to the discrete sampling distribution of S, a correction for continuity should be applied. For the data in Table 5.6, the standard normal deviate with continuity correction applied is

$$z = \frac{|S| - 1}{\sqrt{n(n-1)(2n+5)/18}} = \frac{40 - 1}{\sqrt{12(12-1)[(2)(12) + 5]/18}} = +2.6743 \ ,$$

yielding an approximate two-tail probability value of $P = 0.0074$, under the assumptions of independence, random sampling, and normality. In this case, the difference between the Monte Carlo and asymptotic probability values is only $P(0.0191 - 0.0074) = 0.0117$.

5.6 Coefficient of Concordance

The measurement of the degree of association among multiple sets of rank scores is useful in studies of inter-test reliability. Whereas Spearman's rank-order correlation coefficient and Kendall's τ_a and τ_b measures express the degree of association between two variables measured in, or transformed to, rank scores, the coefficient of concordance expresses the degree of association among multiple sets of rank scores.

In 1939, Kendall and Babington Smith developed the well-known coefficient of concordance [4]. Let n and m denote the number of rank scores and the number of judges, respectively, and then Kendall and Babington Smith defined the coefficient of concordance as

$$W = \frac{12S}{m^2(n^3 - n)} , \tag{5.1}$$

where S is the sum of squares of the deviations of sums of ranks from the mean value $m(n + 1)/2$.

Since $m^2(n^3 - n)$ in the denominator of Eq. (5.1) is invariant over all permutations of the data, it is only necessary to consider the distribution of S by arranging the n ranks in all possible ways. Letting one of the rankings be fixed, there are $(n!)^{m-1}$ possible values of S.

In a form more conducive to calculation, W can be defined as

$$W = \frac{12 \sum_{i=1}^{N} R_i^2 - 3m^2 n(n + 1)}{m^2 n(n^2 - 1)} ,$$

where R_i for $i = 1, \ldots, n$ is the sum of the rank scores for the ith of N objects and there are no tied rank scores. It is also well known that W can be defined as a function of the average value of all pairwise Spearman rank-order correlation coefficients given by

$$\bar{\rho} = \frac{2}{m(m - 1)} \sum_{i=1}^{m-1} \sum_{j=i+1}^{m} \rho_{ij} .$$

In this regard, Kendall and Babington Smith showed that $\bar{\rho}$ is simply the intraclass coefficient, r_{I}, for the m sets of rankings. The relationships between W and $\bar{\rho}$ are

$$\bar{\rho} = \frac{mW - 1}{m - 1} \quad \text{and} \quad W = \frac{\bar{\rho}(m - 1) + 1}{m} ,$$

and the relationships between W and r_{I} are

$$r_{\text{I}} = \frac{mW - 1}{m - 1} \quad \text{and} \quad W = \frac{r_{\text{I}}(m - 1) + 1}{m} .$$

If all arrangements of the m sets of rank scores occur with equal chance, the exact probability of W computed on M possible arrangements of the rank scores under the null hypothesis is

$$P(W \geq W_o \,|\, H_0) = \frac{\text{number of } W \text{ values } \geq W_o}{M} ,$$

where W_o is the observed value of W.

5.6.1 Example Analysis

For an example analysis employing Kendall and Babington's Smith's coefficient of concordance, consider the rank data in Table 5.8 with $n = 8$ objects and $m = 3$ sets of rankings. The number of observations is kept small for illustration purposes.

For the rank scores in Table 5.8, the sum of the squared rank scores is

$$\sum_{i=1}^{N} R_i^2 = 11^2 + 15^2 + 4^2 + 6^2 + 16^2 + 132 + 22^2 + 21^2 = 1{,}748 ,$$

Kendall's and Babington Smith's coefficient of concordance is

$$W = \frac{12\sum_{i=1}^{n} R_i^2 - 3m^2n(n+1)}{m^2n(n^2-1)}$$

$$= \frac{12(1{,}748) - (3)(3^2)(8)(8+1)^2}{3^2(8)(8^2-1)} = 0.7672 ,$$

Table 5.8 Data for Kendall and Babington Smith's coefficient of concordance with $m = 3$ sets of rankings and $n = 8$ objects

	Ranking			
Object	1	2	3	R_i
1	4	2	5	11
2	6	6	3	15
3	2	1	1	4
4	1	3	2	6
5	3	5	8	16
6	5	4	4	13
7	7	8	7	22
8	8	7	6	21

and Spearman's pairwise average rank-order correlation with

$$\rho_{12} = 1 - \frac{6\sum_{i=1}^{N} d_i^2}{N(N^2-1)} = 1 - \frac{6(16)}{8(8^2-1)} = +0.8095 ,$$

$$\rho_{13} = 1 - \frac{6\sum_{i=1}^{N} d_i^2}{N(N^2-1)} = 1 - \frac{6(42)}{8(8^2-1)} = +0.50 ,$$

and

$$\rho_{23} = 1 - \frac{6\sum_{i=1}^{N} d_i^2}{N(N^2-1)} = 1 - \frac{6(30)}{8(8^2-1)} = +0.6429$$

is

$$\bar{\rho} = \frac{2}{m(m-1)} \sum_{i=1}^{m-1} \sum_{j=i+1}^{m} \rho_{ij} = \frac{2(0.8095 + 0.50 + 0.6429)}{3(3-1)} = 0.6508 .$$

Finally, the relationships between $\bar{\rho}$ and W are

$$\bar{\rho} = \frac{mW - 1}{m - 1} = \frac{3(0.7672) - 1}{3 - 1} = 0.6508$$

and

$$W = \frac{\bar{\rho}(m-1) + 1}{m} = \frac{(0.6508)(3-1) + 1}{3} = 0.7672 .$$

For the data in Table 5.8, the grand mean is

$$\bar{\bar{x}} = \frac{1}{nk} \sum_{i=1}^{nk} x_i = \frac{4 + 2 + 5 + \cdots + 8 + 7 + 6}{(8)(3)} = \frac{108}{24} = 4.50 ,$$

the sample variance is

$$S_x^2 = \frac{1}{nk}\sum_{i=1}^{nk}\left(x_i - \bar{\bar{x}}\right)^2$$

$$= \frac{(4-4.50)^2 + (2-4.50)^2 + (54.50)^2 + \cdots + (6-4.50)^2}{(8)(3)}$$

$$= \frac{126}{24} = 5.25\ ,$$

the sum of squares is calculated in Table 5.9, and the intraclass correlation coefficient is

$$r_{\text{I}} = \frac{m\sum_{i=1}^{n}\left(\bar{x}_i - \bar{\bar{x}}\right)^2}{n(m-1)S_x^2} - \frac{1}{m-1} = \frac{(3)(32.2222)}{8(3-1)(5.25)} - \frac{1}{3-1} = +0.6508\ .$$

Thus it is demonstrated that Spearman's $\bar{\rho}$ and the intraclass correlation coefficient are equivalent. It follows that the intraclass correlation coefficient, defined in terms of Kendall and Babington Smith's W, is

$$r_{\text{I}} = \frac{mW-1}{m-1} = \frac{3(0.7672)-1}{3-1} = 0.6508,$$

and Kendall and Babington Smith's W, defined in terms of the intraclass correlation coefficient, is

$$W = \frac{r_{\text{I}}(m-1)+1}{m} = \frac{(0.6508)(3-1)+1}{3} = 0.7672\ .$$

Table 5.9 Means, grand mean, and sum of squares for the data in Table 5.8

Object	x_1	x_2	x_3	$\bar{x}$	$\bar{\bar{x}}$	$(\bar{x}-\bar{\bar{x}})^2$
1	4	2	5	3.6667	4.50	0.6944
2	6	6	3	5.0000	4.50	0.2500
3	2	1	1	1.3333	4.50	10.0278
4	1	3	2	2.0000	4.50	6.2500
5	3	5	8	5.3333	4.50	0.6944
6	5	4	4	4.3333	4.50	0.0278
7	7	8	7	7.3333	4.50	8.0278
8	8	7	6	7.0000	4.50	6.2500
Sum						32.2222

5.6.2 Permutation Probability

For the $m = 3$ sets of rank scores in Table 5.7, there are

$$M = \left(n!\right)^{m-1} = \left(8!\right)^{3-1} = 1{,}625{,}702{,}400$$

possible arrangements of the data.While the number of possible arrangements makes an exact permutation analysis impractical, it is not impossible. If the reference set of all arrangements of the rank scores in Table 5.7 occur with equal chance, the exact probability of $W = 0.7672$ under the null hypothesis is

$$P(W \geq W_o \mid H_0) = \frac{\text{number of } W \text{ values } \geq W_o}{M}$$

$$= \frac{5{,}125{,}594}{1{,}625{,}702{,}400} = 3.1528 \times 10^{-3} \, ,$$

where W_o is the observed value of W.

5.6.3 Asymptotic Probability

It is easily shown that Kendall's S statistic can be defined in terms of Kendall and Babington Smith's W. Given

$$W = \frac{12S}{m^2(n^3 - n)} \quad \text{then} \quad S = \frac{Wm^2(n^3 - n)}{12} \, .$$

For the rank-score data in Table 5.7 with $n = 8$, $m = 3$, and $W = 0.7672$, Kendall's S is

$$S = \frac{Wm^2(n^3 - n)}{12} = \frac{(0.7672)(3^2)(8^3 - 8)}{12} = 29 \, .$$

Kendall's S is distributed $N(0, 1)$ under the null hypothesis with a mean of zero and variance

$$\sigma_S^2 = \frac{n(n-1)(2n+5)}{18}$$

as $n \to \infty$. Since the normal distribution is an approximation to the discrete sampling distribution of S, a correction for continuity should be applied. For the data in Table 5.7, the standard normal deviate with continuity correction applied is

$$z = \frac{|S| - 1}{\sqrt{n(n-1)(2n+5)/18}} = \frac{29 - 1}{\sqrt{8(8-1)[(2)(8)+5]/18}} = +3.4641 \ ,$$

yielding an approximate two-tail probability value of $P = 2.6600 \times 10^{-4}$, under the assumptions of independence, random sampling, and normality. For this example analysis, the difference between the exact and asymptotic probability values is only $P(3.1528 \times 10^{-3} - 2.6600 \times 10^{-4}) = 2.8868 \times 10^{-3}$.

5.7 Summary

Chapter 5 presented permutation statistical methods, both exact and Monte Carlo, as applied to two cross-classified interval-level variables. The measures of relationship described in the chapter included Pearson's product-moment correlation coefficient, Fisher's intraclass correlation coefficient, Spearman's rank-order correlation coefficient, Spearman's footrule measure, and Kendall and Babington Smith's coefficient of concordance. Relationships between and among the various measures were described and illustrated with example analyses.

References

1. K.J. Berry, P.W. Mielke, Spearman's footrule as a measure of agreement. Psychol. Rep. **80**, 839–846 (1997)
2. E.A. Haggard, *Intraclass Correlation and the Analysis of Variance* (Dryden, New York, 1958)
3. M.G. Kendall, *Rank Correlation Methods*, 3rd edn. (Griffin, London, 1962)
4. M.G. Kendall, B. Babington Smith, The problem of m rankings. Ann. Math. Stat. **10**, 275–287 (1939)
5. K.O. McGraw, S.P. Wong, Forming inferences about some intraclass correlation coefficients. Psychol. Meth. **1**, 30–46 (1996)
6. R. Sitgreaves, Review of "Intraclass Correlation and the Analysis of Variance" by E.A. Haggard. J. Am. Stat. Assoc. **55**, 384–385 (1960)
7. P.E. Shrout, J.L. Fleiss, Intraclass correlations: uses in assessing rater reliability. Psychol. Bull. **86**, 420–428 (1979)
8. C.E. Spearman, The proof and measurement of association between two things. Am. J. Psychol. **15**, 72–101 (1904)
9. C.E. Spearman, 'Footrule' for measuring correlation. Brit. J. Psychol. **2**, 89–108 (1906)
10. R.M. Vogol, S.M. Papalexiou, J.R. Lamontague, F.C. Dolan, When heavy tails disrupt statistical inference. Am. Stat. **79**, 221–235 (2025)
11. S.P. Wong, K.O. McGraw, Confidence intervals and F tests for intraclass correlations based on three-way random effects models. Educ. Psychol. Meas. **59**, 270–288 (1999)

Chapter 6
Mixed Measures of Association

Abstract This chapter describes measures of association for two variables at different levels of measurement, e.g., a nominal-independent variable and an ordinal- or interval-dependent variable, and an ordinal-independent variable and an interval dependent variable. Included in the chapter are Freeman's measure of nominal–ordinal association (θ), Cureton's rank–biserial correlation coefficient (r_{rb}), the point–biserial correlation coefficient (r_{pb}), the biserial correlation coefficient (r_b), and Jaspen's multiserial correlation coefficient ($r_{y\bar{z}}$).

Keywords Point-biserial correlation · Biserial correlation · Multiserial correlation · Jaspen's test

6.1 Introduction

In this chapter, exact and Monte Carlo permutation statistical methods are applied to measures of association designed for two variables at different levels of measurement, e.g., a nominal-independent variable and an ordinal- or interval dependent variable, and an ordinal independent variable and an interval dependent variable. For practical use, there is little reason to examine those cases in which the dependent variable is at a lower measurement level than the independent variable because the underlying logic of prediction cannot use the added information contained within the independent variable [6, p. 292].

Five sections describe measures of association between nominal and ordinal variables, nominal and interval variables, and ordinal and interval variables. Section 6.2 presents permutation statistical methods for measuring association between a nominal-independent variable and an ordinal-dependent variable. Section 6.3 presents permutation statistical methods for measuring the association between a dichotomous nominal variable and an ordinal variable. Section 6.4 continues the chapter with a discussion of measures of association for a nominal variable and an interval variable. Section 6.5 concludes the chapter with a discussion of permutation statistical methods for an ordinal variable and an interval variable.

J. E. Johnston et al., *Permutation Statistical Methods: Measures of Relationship*, SpringerBriefs in Statistics, https://doi.org/10.1007/978-3-032-26608-8_6

In general, for any permutation analysis of cross-classified data, there are four essential components:

1. The permutation test statistic, δ
2. The probability value, either exact or Monte Carlo, P
3. The exact expected value of test statistic δ, μ_δ
4. The chance-corrected measure of effect size, $\Re$.

6.2 Nominal-Ordinal Association

In 1965, Freeman proposed a measure of association for a nominal-independent variable and an ordinal-dependent variable called theta (θ) [2, pp. 108–119]. Consider an $r \times c$ contingency table where the r rows are a nominal-independent variable (x) and the c columns are an ordinal-dependent variable (y). Let $n_{i.}$, $n_{.j}$, and n_{ij} denote the row marginal frequency totals, column marginal frequency totals, and number of objects in the *ij*th cell, respectively, for $i = 1, \ldots, r$ and $j = 1, \ldots, c$, and let n denote the total number of objects in the contingency table, i.e.,

$$n_{i.} = \sum_{j=1}^{c} n_{ij} \,, \quad n_{.j} = \sum_{i=1}^{r} n_{ij} \,, \quad \text{and} \quad n = \sum_{i=1}^{r} \sum_{j=1}^{c} n_{ij} \,.$$

For Freeman's θ, it is necessary to calculate the absolute sum of the number of concordant pairs and the number of discordant pairs for all combinations of the nominal-independent variable considered two at a time. Thus, assuming that the ordered variable (y) is underlying continuous (i.e., not discrete) and that ties in ranking simply result from a crude classification on that variable [2, p. 113], Freeman's nominal–ordinal measure of association is

$$\theta = \frac{\displaystyle\sum_{i=1}^{r-1} \sum_{j=i+1}^{r} \left| C_{ij} - D_{ij} \right|}{C + D + T_y} \,,$$

where C is the number of concordant pairs, D is the number of discordant pairs, and T_y is the number of pairs tied on variable y.

6.2.1 *Example Analysis*

For an example analysis employing Freeman's θ, consider the data in Table 6.1 with $r = 4$ disjoint unordered categories (Marital Status) and $c = 5$ disjoint ordered categories (Social Adjustment). When the number of unordered categories is greater

Table 6.1 Data for Freeman's θ with $n = 40$ subjects cross-classified into $r = 4$ categories (x) and $c = 5$ ranks (y)

	Social Adjustment					
Status	5	4	3	2	1	Total
Single	1	2	5	2	0	10
Married	10	5	5	0	0	20
Widowed	0	0	2	2	1	5
Divorced	0	0	0	2	3	5
Total	11	7	12	6	4	40

Source Table 6.1 is adapted from Freeman, *Elementary Applied Statistics* [2, p. 114]

than two, the number of concordant pairs (C) and the number of discordant pairs (D) must be calculated for all $r(r-1)/2$ combinations of row categories. For the data in Table 6.1, the number of concordant pairs is

$$C = \sum_{i=1}^{r-1} \sum_{j=1}^{c-1} x_{ij} \left(\sum_{k=i+1}^{r} \sum_{l=j+1}^{c} x_{kl} \right)$$

$$= (1)(5+5+0+0+0+2+2+1+0+0+2+3)$$

$$+ (2)(5+0+0+2+2+1+0+2+3)$$

$$+ \cdots + (2)(2+3) + (2)(3) = 304 \;,$$

the number of discordant pairs is

$$D = \sum_{i=1}^{r-1} \sum_{j=1}^{c-1} x_{i,c-j+1} \left(\sum_{k=i+1}^{r} \sum_{l=1}^{c-j} x_{kl} \right)$$

$$= (0)(2+0+0+0+2+2+0+0+0+5+5+10)$$

$$+ (2)(0+0+0+2+0+0+5+5+10)$$

$$+ \cdots + (2)(0+0) + (0)(0) = 141 \;,$$

the number of concordant pairs (C), the number of discordant pairs (D), and the number of pairs tied on variable y (T_y) is

$$C + D + T_y = \sum_{i=1}^{r-1} \sum_{j=i+1}^{r} n_{i.}n_{j.} = (10)(20) + (10)(5) + (10)(5)$$

$$+ (20)(5) + (20)(5) + (5)(5) = 525 \;,$$

the concordant and discordant pairs for the $r = 4$ rows considered two at a time are

$$C_{12} = (1)(5+5+0+0) + (2)(5+0+0) + (5)(0+0) + (2)(0) = 20 \ ,$$
$$D_{12} = (0)(10+5+5+0) + (2)(10+5+5) + (5)(10+5) + (2)(10) = 135 \ ,$$

$$C_{13} = (1)(0+2+2+1) + (2)(2+2+1) + (5)(2+1) + (2)(1) = 32 \ ,$$
$$D_{13} = (0)(0+0+2+2) + (2)(0+0+2) + (5)(0+0) + (2)(0) = 4 \ ,$$

$$C_{14} = (1)(0+0+2+3) + (2)(0+2+3) + (5)(2+3) + (2)(3) = 46 \ ,$$
$$D_{14} = (0)(0+0+0+2) + (2)(0+0+0) + (5)(0+0) + (2)(0) = 0 \ ,$$

$$C_{23} = (10)(0+2+2+1) + (5)(2+2+1) + (5)(2+1) + (0)(1) = 90 \ ,$$
$$D_{23} = (0)(0+0+2+2) + (0)(0+0+2) + (5)(0+0) + (5)(0) = 0 \ ,$$

$$C_{24} = (10)(0+0+2+3) + (5)(0+2+3) + (5)(2+3) + (0)(3) = 100 \ ,$$
$$D_{24} = (0)(0+0+0+2) + (0)(0+0+0) + (5)(0+0) + (5)(0) = 0 \ ,$$

$$C_{34} = (0)(0+0+2+3) + (0)(0+2+3) + (2)(2+3) + (2)(3) = 16 \ ,$$
$$D_{34} = (1)(0+0+0+2) + (2)(0+0+0) + (2)(0+0) + (0)(0) = 2 \ ,$$

and Freeman's θ is

$$\theta = \frac{\displaystyle\sum_{i=1}^{r-1} \sum_{j=i+1}^{r} \left| C_{ij} - D_{ij} \right|}{C + D + T_y}$$
$$= \frac{|20 - 135| + |32 - 4| + \cdots + |100 - 0| + |16 - 2|}{525} = 0.7486 \ .$$

6.2.2 *Permutation Probability*

The standard error of Freeman's θ is unknown, so a permutation analysis is essential for obtaining the probability. Usually, a Monte Carlo permutation analysis is recommended for analyzing contingency tables as the number of possible arrangements of cell frequencies is often very large. However, in this case with $n = 40$ subjects, there are only $M = 6{,}340{,}588$ arrangements in the reference set of all permutations of the cell frequencies, making an exact analysis feasible.

If all arrangements occur with equal chance, the exact probability value of Freeman's θ under the null hypothesis is the sum of the hypergeometric point probability values associated with arrangements of cell frequencies with values of θ equal to or greater than $\theta = 0.7486$. Based on the underlying hypergeometric probability distribution, the exact probability value of $\theta = 0.7486$ is $P = 0.2105 \times 10^{-10}$.

6.3 Dichotomous Nominal Variables

As a special case of nominal–ordinal association, consider Cureton's rank–biserial correlation, which examines the relationship between a dichotomous variable and an ordinal variable, where, in this case, the dichotomous variable is a nominal variable with only two disjoint unordered categories [1].

For an example analysis of Cureton's rank–biserial correlation, consider an example dataset such as in Table 6.2 in which $n = 10$ objects are ranked (variable y) and also classified into two dummy (0, 1) coded samples (variable x). Cureton defined r_{rb} as

$$r_{rb} = \frac{S}{S_{\max}} ,$$

where S is Kendall's statistic [5] and $S_{\max} = n_0 n_1$, where n_0 is the number of objects coded 0 and n_1 is the number of objects coded 1.

Table 6.2 Example (0, 1) coded data for Cureton's rank–biserial correlation coefficient

	Variable	
Object	x	y
1	0	1
2	1	2
3	0	3
4	0	4
5	0	5
6	0	6
7	1	7
8	0	8
9	1	9
10	1	10

Source Table 6.2 is adapted from Glass, Note on rank–biserial correlation [3]

Table 6.3 Paired differences and concordant (C) and discordant (D) values for the rank scores in Table 6.2

Pair	$x_i - x_j$	$y_i - y_j$	Type	Pair	$x_i - x_j$	$y_i - y_j$	Type
1	$1-0$	$1-2$	C	24	$0-1$	$3-10$	C
2	$0-0$	$1-3$		25	$0-0$	$4-5$	
3	$0-0$	$1-4$		26	$0-0$	$4-6$	
4	$0-0$	$1-5$		27	$0-1$	$4-7$	C
5	$0-0$	$1-6$		28	$0-0$	$4-8$	
6	$0-1$	$1-7$	C	29	$0-1$	$4-9$	C
7	$0-0$	$1-8$		30	$0-1$	$4-10$	C
8	$0-1$	$1-9$	C	31	$0-0$	$5-6$	
9	$0-1$	$1-10$	C	32	$0-1$	$5-7$	C
10	$1-0$	$2-3$	D	33	$0-0$	$5-8$	
11	$1-0$	$2-4$	D	34	$0-1$	$5-9$	C
12	$1-0$	$2-5$	D	35	$0-1$	$5-10$	C
13	$1-0$	$2-6$	D	36	$0-1$	$6-7$	C
14	$1-1$	$2-7$		37	$0-0$	$6-8$	
15	$1-0$	$2-8$	D	38	$0-1$	$6-9$	C
16	$1-1$	$2-9$		39	$0-1$	$6-10$	C
17	$1-1$	$2-10$		40	$1-0$	$7-8$	D
18	$0-0$	$3-4$		41	$1-1$	$7-9$	
19	$0-0$	$3-5$		42	$1-1$	$7-10$	
20	$0-0$	$3-6$		43	$0-1$	$8-9$	C
21	$0-1$	$3-7$	C	44	$0-1$	$8-10$	C
22	$0-0$	$3-8$		45	$1-1$	$9-10$	
23	$0-1$	$3-9$	C				

Table 6.3 lists the

$$\binom{n}{2} = \frac{n(n-1)}{2} = \frac{10(10-1)}{2} = 45$$

possible paired comparisons of x_i and x_j with y_i and y_j, where $i < j$. Each paired difference is labeled as concordant (C) or discordant (D). Paired differences not labeled as C or D are irrelevant in the present context as they are tied by either $x_i = x_j = 0$ or $x_i = x_j = 1$. In Table 6.3, there are $C = 18$ concordant and $D = 6$ discordant paired differences. Thus, for the paired differences in Table 6.3, Kendall's S is $C - D = 18 - 6 = +12$ and Cureton's rank–biserial coefficient is

$$r_{rb} = \frac{S}{n_0 n_1} = \frac{+12}{(6)(4)} = +0.50 \;.$$

6.3.1 Permutation Probability

For the rank scores in Table 6.2, there are only

$$M = \frac{n!}{n_0!\, n_1!} = \frac{10!}{6!\, 4!} = 210$$

arrangements in the reference set of all permutations of the observed scores with $n_0 = 6$ and $n_1 = 4$ preserved for each arrangement, making an exact permutation analysis possible. There are 54 r_{rb} values equal to or greater than $r_{rb} = +0.50$. If all arrangements of the $n = 10$ rank scores occur with equal chance, the exact probability of $r_{rb} = +0.50$ computed on the M possible arrangements of the observed data is

$$P(r_{rb} \geq r_o \,|\, \text{H}_0) = \frac{\text{number of } r_{rb} \text{ values } \geq r_o}{M} = \frac{54}{210} = 0.2571 \,,$$

where r_o is the observed value of Cureton's r_{rb}.

6.3.2 Asymptotic Probability

In testing the significance of Cureton's r_{rb}, it is efficient to apply a test directly to the numerator, S, which is asymptotically distributed $N(0, 1)$ under the null hypothesis with a mean of zero and variance

$$\sigma_S^2 = \frac{n(n-1)(2n+5)}{18}$$

as $n \to \infty$. Since the normal distribution is an approximation to the discrete sampling distribution of S, a correction for continuity should be applied. For the data in Table 6.2, the standard normal deviate with continuity correction applied is

$$z = \frac{|S| - 1}{\sqrt{n(n-1)(2n+5)/18}} = \frac{12 - 1}{\sqrt{10(10-1)[(2)(10)+5]/18}} = +0.9839 \,,$$

yielding an approximate two-tail probability value of $P = 0.3252$, under the assumptions of independence, random sampling, and normality. For this example analysis, the difference between the exact and asymptotic probability values is $P(0.3252 - 0.2571) = 0.0681$.

6.4 Nominal-Interval Association

As special cases of nominal-interval association, consider two measures: point-biserial correlation and biserial correlation. Both examine the relationship between a dichotomous variable and an interval variable where, in this case, the dichotomous variable is a nominal variable with two disjoint unordered categories.

6.4.1 Point-Biserial Correlation

In this section, a measure of association between a dichotomous variable and an interval variable is described. Point-biserial correlation measures the association between a true dichotomous variable and an interval variable and is an important measure in education and educational psychology where it is typically used to measure the correlation between test questions scored as correct (1) or incorrect (0) and the overall score on tests for n students. A low or negative point-biserial correlation coefficient indicates that the students with the highest scores on the test answered the question incorrectly and the students with the lowest scores on the test answered the question correctly, alerting the instructor that the question might be faulty.

6.4.2 Example Analysis

For an example analysis employing the point-biserial correlation coefficient, consider the dichotomous data in Table 6.4 for $n = 20$ objects where variable x is the

Table 6.4 Dummy (0, 1) coded data for the point-biserial correlation coefficient

	Variable			Variable	
Object	x	y	Object	x	y
1	0	99	11	1	86
2	0	99	12	1	90
3	1	98	13	0	97
4	1	98	14	0	95
5	1	97	15	1	92
6	0	89	16	0	98
7	0	95	17	1	86
8	0	94	18	1	85
9	1	92	19	0	94
10	1	60	20	0	96

dichotomous variable and variable y is an unspecified interval variable. The point-biserial correlation is usually computed as

$$r_{pb} = \frac{\bar{y}_1 - \bar{y}_0}{s_y} \sqrt{\frac{n_0 n_1}{n(n-1)}} ,$$

where n_0 and n_1 denote the number of y values coded 0 and 1, respectively, $n = n_0 + n_1$, $\bar{y}_0$ and $\bar{y}_1$ denote the means of the y values coded 0 and 1, respectively, and s_y is the sample standard deviation of the y values given by

$$s_y = \sqrt{\frac{1}{n-1} \sum_{i=1}^{n} (y_i - \bar{y})^2} .$$

For the data in Table 6.4, $n_0 = n_1 = 10$, the mean of the y values coded 0 is

$$\bar{y}_0 = \frac{1}{n_0} \sum_{i=1}^{n_0} y_i = \frac{99 + 99 + \cdots + 94 + 96}{10} = 88.40 ,$$

the mean of the y values coded 1 is

$$\bar{y}_1 = \frac{1}{n_1} \sum_{i=1}^{n_1} y_i = \frac{98 + 98 + \cdots + 86 + 85}{10} = 95.60 ,$$

the mean of the y values coded 0 and 1 (i.e., all the y values) is

$$\bar{y} = \frac{1}{n} \sum_{i=1}^{n} = \frac{99 + 99 + \cdots + 94 + 96}{20} = 92.00 ,$$

the sum of squares of the y values is

$$\sum_{i=1}^{n} (y_i - \bar{y})^2 = 1{,}456 \text{ , as illustrated in Table 6.5,}$$

the standard deviation of the y values is

$$s_y = \sqrt{\frac{1}{n-1} \sum_{i=1}^{n} (y_i - \bar{y})^2} = \sqrt{\frac{1{,}456}{20-1}} = 8.7539 ,$$

Table 6.5 Deviation and squared deviation scores for variable y

y	$y - \bar{y}$	$(y - \bar{y})^2$	y	$y - \bar{y}$	$(y - \bar{y})^2$
99	+7	49	86	−6	36
99	+7	49	90	+2	4
98	+6	36	97	+5	25
98	+6	36	95	+3	9
97	+5	25	92	0	0
89	−3	9	98	+6	36
95	−3	9	86	−6	36
94	+2	4	85	−7	49
92	0	0	94	+2	4
60	−32	1024	96	+4	16
Sum					1456

and the point-biserial correlation coefficient is

$$r_{pb} = \frac{\bar{y}_1 - \bar{y}_0}{s_y}\sqrt{\frac{n_0 n_1}{n(n-1)}} = \frac{95.60 - 88.40}{8.7539}\sqrt{\frac{(10)(10)}{20(20-1)}} = +0.4219 \; .$$

Note that r_{pb} can be calculated simply as the Pearson product-moment correlation (r_{xy}) between dichotomous variable x and interval variable y, although the sign might differ due to the assignment of the 0s and 1s. For the data in Table 6.4

$$\sum_{i=1}^{N} x_i = 1{,}840, \; \sum_{i=1}^{N} y_i = 10, \; \sum_{i=1}^{N} x_i^2 = 170{,}736, \; \sum_{i=1}^{N} y_i^2 = 10, \; \sum_{i=1}^{N} x_i y_i = 884 \; ,$$

and Pearson's product-moment correlation coefficient is

$$r_{xy} = \frac{n\sum_{i=1}^{n} x_i y_i - \sum_{i=1}^{n} x_i \sum_{i=1}^{n} y_i}{\sqrt{\left[n\sum_{i=1}^{n} x_i^2 - \left(\sum_{i=1}^{n} x_i\right)^2\right]\left[n\sum_{i=1}^{n} y_i^2 - \left(\sum_{i=1}^{n} y_i\right)^2\right]}}$$

$$= \frac{(20)(884) - (1{,}840)(10)}{\sqrt{[(20)(170{,}736) - (1{,}840)^2][(20)(10) - (10)^2]}} = -0.4218 \; .$$

However, using this approach, there are

$$M = n! = 20! = 2{,}432{,}902{,}008{,}176{,}640{,}000$$

Table 6.6 Data from Table 6.4 with the y values arranged in two groups of $n_0 = n_1 = 10$

Group	Score									
0	99	99	98	97	96	95	95	94	94	89
1	98	98	97	92	92	92	90	86	86	60

arrangements of the data to be considered. A more efficient approach is to consider the data in Table 6.4 as two groups of observations, as in Table 6.6, and compute the difference between the means of the groups. It is not necessary that $n_0 = n_1$.

6.4.3 Permutation Probability

For the grouped scores in Table 6.6, there are only

$$M = \frac{n!}{n_0!\, n_1!} = \frac{20!}{10!\; 10!} = 184{,}756$$

arrangements in the reference set of all permutations of the observed scores, making an exact permutation analysis possible. There are 5,648 r_{pb} values equal to or greater than $r_{pb} = +0.4219$. If all arrangements of the $n = 20$ scores occur with equal chance, the exact probability value of $r_{pb} = +0.4219$ computed on the M possible arrangements of the observed data with $n_0 = n_1 = 10$ preserved for each arrangement is

$$P(r_{pb} \geq r_\text{o} \mid \text{H}_0) = \frac{\text{number of } r_{pb} \text{ values } \geq r_\text{o}}{M} = \frac{5{,}648}{184{,}756} = 0.0306\;,$$

where r_o is the observed value of r_{pb}.

6.4.4 Asymptotic Probability

The point-biserial correlation coefficient is distributed as Student's t with $n - 2$ degrees of freedom. For the data in Table 6.4,

$$t = r_{pb}\sqrt{\frac{n-2}{1-r_{pb}^2}} = +0.4219\sqrt{\frac{20-2}{1-(0.4219)^2}} = 1.9743\;,$$

yielding a two-tail probability value of $P = 0.0484$. For this example analysis, the difference between the exact and asymptotic probability values is only $P(0.0484 - 0.0306) = 0.0178$.

6.4.5 Biserial Correlation

Whereas point-biserial correlation measures the association between an interval variable and a dichotomous variable that is a true dichotomy, such as correct or incorrect, biserial correlation measures the association between an interval variable and a variable assumed to be continuous, but has been dichotomized, such as IQ dichotomized into "at or below 100" and "above 100" or height dichotomized into "at or below 70 inches" and "above 70 inches." The biserial correlation coefficient is

$$r_b = \frac{(\bar{y}_1 - \bar{y}_0)pq}{uS_y} ,$$

where p and q denote the proportions of all y values coded 0 and 1, respectively, $\bar{y}_0$ and $\bar{y}_1$ denote the means of the y values coded 0 and 1, respectively, S_y is the standard deviation of the y values given by[1]

$$S_y = \sqrt{\frac{1}{n}\sum_{i=1}^{n}(y_i - \bar{y})^2} ,$$

and u is the ordinate of the unit normal distribution at the point of division between the p and q proportions under the distribution given by

$$u = \frac{\exp(-z^2/2)}{\sqrt{2\pi}} .$$

6.4.6 Example Analysis

For an example analysis employing the biserial correlation coefficient, consider the data in Table 6.7 in which $n = 15$ subjects are scored on Work Effectiveness (interval variable y) and are classified into Type A (0) and Type B (1) personalities (dichotomous variable x). For the data in Table 6.7, $n_0 = 6$, $n_1 = 9$, $p = n_0/n = 6/15 = 0.40$, $q = n_1/n = 9/15 = 0.60$, the mean of the y values coded 0 is

$$\bar{y}_0 = \frac{1}{n_0}\sum_{i=1}^{n_0} y_i = \frac{12 + 15 + \cdots + 13 + 11}{6} = 13.3333 ,$$

[1] Note that in this case, the sum of squared deviations is divided by n, not $n - 1$.

Table 6.7 Biserial correlation data on $n = 15$ objects

Object	Type	Effectiveness
1	0	12
2	0	15
3	0	11
4	0	18
5	0	13
6	0	11
7	1	10
8	1	33
9	1	19
10	1	21
11	1	29
12	1	12
13	1	19
14	1	23
15	1	16

the mean of the y values coded 1 is

$$\bar{y}_1 = \frac{1}{n_1}\sum_{i=1}^{n_1} y_i = \frac{10 + 33 + \cdots + 23 + 16}{9} = 20.2222 \; ,$$

the mean of the y values coded 0 and 1 (i.e., all the y values) is

$$\bar{y} = \frac{1}{n}\sum_{i=1}^{n} = \frac{12 + 15 + \cdots + 23 + 16}{15} = 17.4667 \; ,$$

the sum of squares of the y values is

$$\sum_{i=1}^{n} (y_i - \bar{y})^2 = 649.7333 \; , \text{ as illustrated in Table 6.8,}$$

the standard deviation of the y values is

$$S_y = \sqrt{\frac{1}{n}\sum_{i=1}^{n} (y_i - \bar{y})^2} = \sqrt{\frac{649.7333}{15}} = 6.5815 \; ,$$

Table 6.8 Deviation and squared deviation scores for variable y

y	$y - \bar{y}$	$(y - \bar{y})^2$
12	−5.4667	29.9844
15	−2.4667	6.0844
11	−6.4667	41.8178
18	+0.5333	0.2844
13	−4.4667	19.9511
11	−6.4667	41.8178
10	−7.4667	55.7511
33	+15.5333	241.2845
19	+1.5333	2.3511
21	+3.5333	12.4844
29	+11.5333	133.0178
12	−5.4667	29.8844
19	+1.5333	2.3511
23	+5.5333	30.6178
16	−1.4667	2.1511
Sum		649.7333

the standard score that defines the lower $p = 0.40$ of the unit normal distribution is $z = -0.2433$, the ordinate of the unit normal distribution is

$$u = \frac{\exp(-z^2/2)}{\sqrt{2\pi}} = \frac{\exp[-(-0.2533)^2/2]}{\sqrt{2(3.1416)}} = 0.3863 \ ,$$

and the biserial correlation coefficient is

$$r_b = \frac{(\bar{y}_1 - \bar{y}_0)pq}{uS_y} = \frac{(20.2222 - 13.3333)(0.40)(0.60)}{(0.3863)(6.5815)} = +0.6503 \ .$$

6.4.7 Permutation Probability

For the data in Table 6.7, there are only

$$M = \frac{n!}{n_0!\, n_1!} = \frac{15!}{6!\, 9!} = 5{,}005$$

arrangements in the reference set of all permutations of the observed scores, making an exact permutation analysis possible. Note that in the equation for the biserial correlation coefficient,

$$r_b = \frac{\bar{y}_1 - \bar{y}_0 pq}{uS_y} \ ,$$

p, q, u, and S_y are invariant under permutation. Therefore, the permutation distribution can efficiently be based entirely on $\bar{y}_1 - \bar{y}_0$. There are 263 r_b values equal to or greater than $r_b = +0.6503$. If all $M = 5{,}005$ arrangements of the $n = 15$ observed values occur with equal chance, the exact probability of $r_b = +0.6503$ computed on the M possible arrangements of the observed data with $n_0 = 6$ and $n_1 = 9$ preserved for each arrangement is

$$P(r_b \geq r_\mathrm{o} \,|\, \mathrm{H}_0) = \frac{\text{number of } r_b \text{ values } \geq r_\mathrm{o}}{M} = \frac{263}{5{,}005} = 0.0525 \ ,$$

where r_o is the observed value of r_b.

6.4.8 Asymptotic Probability

The biserial correlation coefficient is distributed as Student's t with $n - 2$ degrees of freedom. For the data in Table 6.6,

$$t = r_b \sqrt{\frac{n-2}{1-r_b^2}} = +0.5128 \sqrt{\frac{15-2}{1-(0.5128)^2}} = +2.1537 \ ,$$

yielding a two-tail probability value of $P = 0.0506$. For this example analysis, the difference between the exact and asymptotic probability values is only $P(0.0525 - 0.0506) = 0.0019$.

6.5 Ordinal-Interval Association

This section describes the relationship between an ordinal variable and an interval variable. Jaspen's multiserial correlation coefficient is simply the Pearson product-moment correlation coefficient between an interval variable, Y, and a transformation of an ordinal variable, X [4]. The procedure assumes that the rank categories of the ordinal variable are based on an underlying normally distributed interval scale. Once the assumption is satisfied, percentile ranks are converted to standard scores through an inverse normal probability function.

For the procedure to be valid, it is necessary to assume that each score is at the mean of the category on the corresponding underlying interval scale. Given n values on the interval variable and k disjoint ordered categories on the ordinal variable, the mean standard score of the underlying scale for a given category is

$$\bar{z}_j = \frac{Y_{L,j} - Y_{U,j}}{p_j} \qquad \text{for } j = 1, \ \ldots, \ k \ ,$$

where $Y_{L,j}$ and $Y_{U,j}$ are the lower and upper ordinates, respectively, of the segment of the $N(0, 1)$ distribution corresponding to the jth ordered category, and where p_j is the proportion of cases in the jth of k categories. Given the obtained values of $\bar{z}_j$, $j = 1, \ldots, k$, and the original n values of the interval variable, Pearson's product-moment correlation between the Y and $\bar{z}$ values yields Jaspen's multiserial correlation.

However, the resulting multiserial correlation coefficient is usually biased. It is possible to estimate, by means of an appropriate transformation, the correlation between the interval variable Y and the continuum hypothesized to underlie the ordinal variable X. The transformation is essentially a correction for grouping. As a result of classifying the underlying interval values into k ordinal categories, the multiserial correlation coefficient obtained from the grouped data will be smaller than the corresponding Pearson product-moment correlation coefficient that would have been obtained had the interval values been available for both variables.

The appropriate transformation to obtain a corrected multiserial correlation coefficient is to divide the obtained multiserial correlation coefficient by the assumed correlation between the underlying interval variable and the midpoints of the k ordinal categories. The assumed correlation is simply the standard deviation of the $\bar{z}$ scores corresponding to the midpoints of the ordinal categories given by

$$S_{\bar{z}} = \left(\frac{1}{n} \sum_{j=1}^{k} n_j \bar{z}_j^2 \right)^{1/2} .$$

6.5.1 Example Analysis

For an example analysis employing Jaspen's multiserial correlation coefficient, consider the data in Table 6.9 in which $n = 32$ interval values (IQ scores) are listed in $k = 4$ ordered categories (Hypnotic Susceptibility).

Table 6.10 illustrates the calculation of Jaspen's multiserial correlation. The first column lists the $k = 4$ ordered categories of variable X. The second column lists the number of observations in each of the k categories. The third column lists the proportion of observations in each of the k categories, e.g., for rank categories 4 and 3,

$$p_4 = \frac{n_4}{n} = \frac{4}{32} = 0.1250 \quad \text{and} \quad p_3 = \frac{n_3}{n} = \frac{10}{32} = 0.3125 .$$

The fourth column lists the cumulative proportion of observations in each of the k categories, e.g., for rank category 3, $P_3 = p_4 + p_3 = 0.1250 + 0.3125 = 0.4375$. The fifth column lists the standard score that defines the cumulative proportion from the fourth column under the unit normal distribution for each of the k categories, e.g., for rank category 4, the standard score that defines the lowest (left tail) of the normal

Table 6.9 Intelligence test scores for $k = 4$ ranks on hypnotic susceptibility

Susceptibility			
4	3	2	1
136	144	139	128
131	137	134	111
126	134	133	104
116	131	132	103
	129	130	103
	126	129	101
	122	123	101
	117	117	
	111	116	
	109	112	
		106	

Source Table 6.9 is adapted from Freeman, *Elementary Applied Statistics* [2, p. 139]

Table 6.10 Calculation of the mean standard scores for the $k = 4$ ordinal categories

X	n	p	P	z	Y_L	Y_U	$\bar{z}$
4	4	0.1250	0.1250	−1.1503	0.2059	0.0000	+1.6472
3	10	0.3125	0.4375	−0.1573	0.3940	0.2059	+0.6022
2	11	0.3438	0.7813	+0.7766	0.2951	0.3940	−0.2880
1	7	0.2188	1.0000	+1.0000	0.0000	0.2951	−1.3487
Total	32	1.0000					

distribution is $z = -1.1503$. The sixth column lists the height of the ordinate at the standard score in the fifth column below the segment in question of the unit–normal distribution for each of the k categories, e.g., for rank category 3,

$$Y_{L,3} = \frac{\exp(-z^2/2)}{\sqrt{2\pi}} = \frac{\exp[-(-0.1573)^2/2]}{\sqrt{2(3.1416)}} = 0.3940\ .$$

The seventh column lists the height of the ordinate at the standard score in the fifth column above the segment in question of the unit normal distribution for each of the k categories, e.g., for rank category 3,

$$Y_{U,3} = \frac{\exp(-z^2/2)}{\sqrt{2\pi}} = \frac{\exp[-(-1.1503)^2/2]}{\sqrt{2(3.1416)}} = 0.2059\ .$$

The last column lists the average standard scores for each of the k categories, e.g., for rank category 4,

$$\bar{z}_4 = \frac{Y_{L,4} - Y_{U,4}}{p_4} = \frac{0.2059 - 0.0000}{0.1250} = +1.6472 \; .$$

The multiserial correlation is the Pearson product-moment correlation between the Y interval values in Table 6.9 and the $\bar{z}$ values in Table 6.10. Table 6.11 lists the Y, $\bar{z}$, Y^2, $\bar{z}^2$, and $Y\bar{z}$ values, along with the corresponding sums. For the summations in Table 6.11, the Pearson product-moment correlation between the Y and $\bar{z}$ values is

$$r_{Y\bar{z}} = \frac{n\sum_{i=1}^{n} Y_i\bar{z}_i - \sum_{i=1}^{n} Y_i \sum_{i=1}^{n} \bar{z}_i}{\sqrt{\left[n\sum_{i=1}^{n} Y_i^2 - \left(\sum_{i=1}^{n} Y_i\right)^2\right]\left[n\sum_{i=1}^{n} \bar{z}_i^2 - \left(\sum_{i=1}^{n} \bar{z}_i\right)^2\right]}}$$

$$= \frac{(32)(189.4757) - (3{,}891)(0.00)}{\sqrt{\left[(32)(478{,}029) - (3{,}891)^2\right]\left[(32)(28.1428) - (0.00)^2\right]}} = +0.5094 \; .$$

Then the correction for grouping is

$$S_{\bar{z}} = \left(\frac{1}{n}\sum_{j=1}^{k} n_j\bar{z}_j^2\right)^{1/2} = \left\{\frac{1}{32}\left[(4)(+1.6472)^2 + (10)(+0.6022)^2 + (11)(-0.2880)^2 + (7)(-1.3487)^2\right]\right\}^{1/2} = 0.9375$$

and the corrected multiserial correlation coefficient is

$$r_c = \frac{r_{Y\bar{z}}}{S_{\bar{z}}} = \frac{+0.5094}{0.9375} = +0.5434 \; .$$

6.5.2 Permutation Probability

Because there are

$$M = n! = 32! = 263{,}130{,}836{,}933{,}693{,}530{,}167{,}218{,}012{,}160{,}000{,}000 \; ,$$

Table 6.11 Calculation of the sums needed for the product-moment correlation between variables Y and $\bar{z}$

Rank	Y	$\bar{z}$	Y^2	$\bar{z}^2$	$Y\bar{z}$
4	136	+1.6472	18,496	2.7133	+224.0192
	131	+1.6472	17,161	2.7133	+215.7832
	126	+1.6472	15,876	2.7133	+207.5472
	116	+1.6472	13,456	2.7133	+191.0752
3	144	+0.6022	20,736	0.3626	+86.7168
	137	+0.6022	18,769	0.3626	+82.5014
	134	+0.6022	17,956	0.3626	+80.6948
	131	+0.6022	17,161	0.3626	+78.8882
	129	+0.6022	16,641	0.3626	+77.6838
	126	+0.6022	15,876	0.3626	+75.8772
	122	+0.6022	14,884	0.3626	+73.4684
	117	+0.6022	13,689	0.3626	+70.4574
	111	+0.6022	12,321	0.3626	+66.8442
	109	+0.6022	11,881	0.3626	+65.6398
2	139	−0.2880	19,321	0.0829	−40.0320
	134	−0.2880	17,956	0.0829	−38.5920
	133	−0.2880	17,689	0.0829	−38.3040
	132	−0.2880	17,424	0.0829	−38.0160
	130	−0.2880	16,900	0.0829	−37.4400
	129	−0.2880	16,641	0.0829	−37.1520
	123	−0.2880	15,129	0.0829	−35.4240
	117	−0.2880	13,689	0.0829	−33.6960
	116	−0.2880	13,456	0.0829	−33.4080
	112	−0.2880	12,544	0.0829	−32.2560
	106	−0.2880	11,236	0.0829	−30.5280
1	128	−1.3487	16,384	1.8190	−172.6336
	111	−1.3487	12,321	1.8190	−149.7057
	104	−1.3487	10,816	1.8190	−140.2648
	103	−1.3487	10,609	1.8190	−138.9161
	103	−1.3487	10,609	1.8190	−138.9161
	101	−1.3487	10,201	1.8190	−136.2187
	101	−1.3487	10,201	1.8190	−136.2187
Sum	3,891	0.0000	478,029	28.1248	+189.4757

possible arrangements of the observed values, an exact permutation analysis is not possible and a Monte Carlo analysis is mandated. There are 3,069 r_c values equal to or greater than $r_c = +0.5434$. If all arrangements of the $n = 20$ scores occur with equal chance, the Monte Carlo probability value of $r_c = +0.5434$ computed on $L = 1{,}000{,}000$ random arrangements of the observed data with $n_1 = 7$, $n_2 = 11$,

$n_3 = 10$, and $n_4 = 4$ preserved for each arrangement is

$$P(r_c \geq r_o \,|\, H_0) = \frac{\text{number of } r_c \text{ values } \geq r_o}{L} = \frac{3{,}069}{1{,}000{,}000} = 3.0690 \times 10^{-3} \, ,$$

where r_o is the observed value of r_c.

6.5.3 *Asymptotic Probability*

In contrast to the permutation probability value, r_c is distributed as Student's t under the null hypothesis with $n - 2$ degrees of freedom. If the population parameter, ρ_c, is assumed to be zero, then for the data in Table 6.7,

$$t = r_c \sqrt{\frac{n-2}{1-r_c^2}} = +0.5434 \sqrt{\frac{32-2}{1-(0.5434)^2}} = +3.5455 \, ,$$

and with $n - 2 = 32 - 2 = 30$ degrees of freedom, the approximate two-sided probability value of $r_c = +0.5434$ is $P = 1.3085 \times 10^{-3}$.

If, as suggested by Jaspen, r_c is converted to a standard score with Fisher's z transform and evaluated with the normal distribution, Fisher's z is

$$z = \frac{\tanh^{-1}(r_c)}{\sqrt{\dfrac{1}{n-3}}} = \frac{\tanh^{-1}(+0.5434)}{\sqrt{\dfrac{1}{32-3}}} = \frac{+0.6090}{0.1857} = +3.2796,$$

yielding an approximate two-sided probability value of $P = 1.0395 \times 10^{-3}$. For this example analysis, the difference between the Monte Carlo and asymptotic probability values is only $P(3.06900 \times 10^{-3} - 1.0395 \times 10^{-3}) = 2.0295 \times 10^{-3}$.

6.6 Summary

Chapter 6 presented permutation statistical methods, both exact and Monte Carlo, as applied to two mixed-level cross-classified variables, such as nominal-ordinal, nominal-interval, and ordinal-interval. The measures of relationship described in the chapter included Freeman's θ for a nominal-independent variable and an ordinal-dependent variable, Cureton's rank-biserial correlation coefficient for a dichotomous nominal-level independent variable and an ordinal-dependent variable, the point-biserial correlation coefficient for a dichotomous nominal-level variable and an interval-level variable, the biserial correlation coefficient for a dichotomous nominal-level variable and an interval-level variable, and Jaspen's multiserial correlation coefficient for an ordinal-level variable and an interval-level variable.

References

1. E.E. Cureton, Rank-biserial correlation. Psychometrika **21**, 287–290 (1956)
2. L.C. Freeman, *Elementary Applied Statistics* (Wiley, New York, 1965)
3. G.V. Glass, Note on rank-biserial correlation. Educ. Psychol. Meas. **26**, 623–631 (1966)
4. N. Jaspen, Serial correlation. Psychometrika **11**, 23–30 (1946)
5. M.G. Kendall, A new measure of rank correlation. Biometrika **30**, 81–93 (1938)
6. R.K. Leik, W.R. Gove, Integrated approach to measuring association, in *Sociological Methodology*, ed. by H.L. Costner (Jossey Bass, San Francisco, CA, 1971), pp. 279–301

Epilogue

This Springer Brief on *Permutation Statistical Methods: Measures of Relationship* is a companion volume to *Permutation Statistical Methods: Tests of Differences* by the same authors. While the previous Brief concentrated on statistical tests of differences, including tests of differences for one-sample tests, two independent samples, two matched samples, multiple independent samples, and multiple matched samples, the current work concentrates on measures of relationship, including measures of nominal-nominal association, ordinal-ordinal association, interval-interval correlation, nominal-ordinal association, nominal-interval association, ordinal-interval association, and agreement.

This book emphasized permutation statistical methods under the Fisher–Pitman model. However, the permutation analyses were always paired with conventional statistical analyses under the Neyman–Pearson model, with which the reader is expected to be familiar. The pairing allowed for comparisons and contrasts between permutation and conventional analyses. Example analyses highlight the differences between the various measures and the permutation and conventional probability values.

Two models of permutation statistical methods were considered: exact and Monte Carlo methods. Exact permutation methods analyze all possible arrangements of the observed data and provide an exact probability value, as the name implies. Monte Carlo permutation methods analyze a large random sample of possible arrangements and provide an approximate, but highly accurate, probability value. The six chapters describe and illustrate permutation applications to various measures of nominal-nominal association, ordinal-ordinal association, correlation and agreement, and mixed measures of association: nominal-ordinal, nominal-interval, and ordinal-interval.

J. E. Johnston et al., *Permutation Statistical Methods: Measures of Relationship*, SpringerBriefs in Statistics, https://doi.org/10.1007/978-3-032-26608-8

Each measure of relationship was described and illustrated with a conventional analysis under the Neyman–Pearson population model and the Fisher–Pitman permutation model. Probability values from both models were calculated and compared for each measure. The assumptions and requirements for each model were delineated and compared for each measure. The authors hope that this Springer Brief will serve as an introduction to permutation statistical methods for the analysis of measures of relationships.

Glossary

ASR	An adjusted standardized residual
C	The number of concordant pairs
D	The number of discordant pairs
E_1	Errors of the first kind
E_2	Errors of the second kind
F	Fisher's F variance-ratio test statistic
FEP	Fisher's exact probability test
H_0	The null hypothesis
H_1	The alternative hypothesis
L	The number of random arrangements
M	The number of possible arrangements
MS	Mean squares
$N(0, 1)$	Normal (Gaussian) distribution
P	A cumulative (tail) probability value
PRE	Proportional reduction in error
R^2	Multiple correlation coefficient
S	Kendall's test statistic
S	Whitfield's test statistic
$S_{\max}$	The maximum value of statistic S
S_x	The sample standard deviation
S_x^2	The sample variance
SR	A standardized residual
SS	The sum of squares
T_x	The number of pairs tied on variable x
T_y	The number of pairs tied on variable y
T_{xy}	The number of pairs tied on variables x and y
UC	Uniform random number for columns
UR	Uniform random number for rows
V^2	Cramér's measure of nominal association

J. E. Johnston et al., *Permutation Statistical Methods: Measures of Relationship*, SpringerBriefs in Statistics, https://doi.org/10.1007/978-3-032-26608-8

W	Coefficient of concordance
df	Degrees of freedom
d_{xy}	Somers' measure of ordinal association, x dependent
d_{yx}	Somers' measure of ordinal association, y dependent
$d_{x.y}$	Kim's measure of ordinal association, x dependent
$d_{y.x}$	Kim's measure of ordinal association, y dependent
e	Wilson's measure of ordinal association
m	Minimum number of rows (r) or columns (c)
n	The sample size
p	A point probability value
r_b	Biserial correlation coefficient
r_{I}	Intraclass correlation coefficient
r_{rb}	Cureton's rank–biserial correlation coefficient
r_{pb}	Point–biserial correlation coefficient
r_{xy}	Pearson's product–moment correlation coefficient
r_{xy}^2	Pearson's coefficient of determination
$r_{Y\bar{z}}$	Jaspen's ordinal–interval correlation coefficient
s_x	An estimated population standard deviation
s_x^2	An estimated population variance
t	Student's t test statistic
t_a	Goodman and Kruskal's nominal measure, a dependent
t_b	Goodman and Kruskal's nominal measure, b dependent
u	Density (height) of the normal distribution
$\bar{x}$	The sample mean
$\bar{\bar{x}}$	The sample grand mean
z	A standard normal deviate
!	The factorial symbol
∞	The infinity symbol
Π	The product symbol
Σ	The summation symbol
$\mathcal{R}$	Spearman's footrule measure
$\Re$	Mielke and Berry's test statistic

Index

J. E. Johnston et al., *Permutation Statistical Methods: Measures of Relationship*, SpringerBriefs in Statistics, https://doi.org/10.1007/978-3-032-26608-8

MIX
Papier aus verantwortungsvollen Quellen
Paper from responsible sources
FSC® C105338

If you have any concerns about our products,
you can contact us on
ProductSafety@springernature.com

In case Publisher is established outside the EU,
the EU authorized representative is:
Springer Nature Customer Service Center GmbH
Europaplatz 3, 69115 Heidelberg, Germany

Printed by Libri Plureos GmbH
in Hamburg, Germany